Lecture Notes in Chemistry

Edited by G. Berthier M. J. S. Dewar H. Fischer
K. Fukui G. G. Hall H. Hartmann H. H. Jaffé J. Jortner
W. Kutzelnigg K. Ruedenberg E. Scrocco

27

Wolfgang Bruns
Ioan Motoc
Kenneth F. O'Driscoll

Monte Carlo Applications in Polymer Science

Springer-Verlag
Berlin Heidelberg New York 1981

Authors

Wolfgang Bruns
Iwan N. Stranski-Institut, Technische Universität Berlin
Ernst-Reuter-Platz 7, 1000 Berlin 10, West Germany

Ioan Motoc
University of Waterloo, Chemical Engineering Department
Waterloo, Ontario N2L 3G1, Canada
(On leave from Chemistry Research
Centre B-dul M. Viteazul 24
1900 Timisoara, Romania)

Kenneth F. O'Driscoll
University of Waterloo, Chemical Engineering Department
Waterloo, Ontario, N2L 3G1, Canada

ISBN-13: 978-3-540-11165-8   e-ISBN-13: 978-3-642-93195-6
DOI: 10.1007/978-3-642-93195-6

Library of Congress Cataloging in Publication Data
Bruns, Wolfgang, 1931-
Monte Carlo applications in polymer science. (Lecture notes in chemistry; 27)
Bibliography: p. Includes index. 1. Polymers and polymerization--Mathematics. 2. Monte
Carlo method. I. Motoc, Ioan, 1950- II. O'Driscoll, Kenneth F. III. Title. IV. Series.
QD381.8.B78   547.7   81-21277 ISBN-13: 978-3-540-11165-8   AACR2

2152/3140-543210

Acknowledgement

One of us (I.M.) is indebted to Professor I. Muscutariu (Timisoara University) for many constructive suggestions concerning the Chapter 1 and to Drs. O. Dragomir-Filimonescu (Institute for Computing Techniques, Timisoara) and R. Vancea (Regional Computer Centre,Suceava) who have written TEST program (Chapter 1) and parts of MEMORY program (Chapter 4), respectively.

CONTENTS

Chapter 1   1

    The Monte Carlo Method and Applications
    I. MOTOC and W. BRUNS

Chapter 2   62

    Monte Carlo Calculation of Sequence Distributions
    in Polymers
    I. MOTOC and F.K. O'DRISCOLL

Chapter 3   105

    Polymer Configuration
    W. BRUNS

Chapter 4   142

    Fortran Programs
    W. BRUNS and I. MOTOC

Chapter 1

THE MONTE CARLO METHOD AND APPLICATIONS

I. Motoc and W. Bruns

1.1. Models

1.2. Random numbers

1.3. Computer Monte Carlo models

   1.3.1. The accuracy of the Monte Carlo Method

   1.3.2. Variance reduction methods

1.4. Monte Carlo applications in Polymer Science

   1.4.1. Solvent around bimacromolecules

   1.4.2. Self-replicating macromolecules

   1.4.3. Molecular weight distribution

   1.4.4. Reactivity in polymer analogous reactions

   1.4.5. Reactivity in binary irreversible copolymerization

   1.4.6. Step-growth polymerizations

   1.4.7. Chain branching and degradation

   1.4.8. Inhomogeneity in copolymers

   1.4.9. Conformation and sequence distribution

In the present chapter, the main concepts used in this book are discussed (i.e., models, random numbers and computer Monte-Carlo models), and then a short review of Monte Carlo applications in Chemistry and an up-to date review of Monte Carlo application in Polymer Science are given. It is hoped that the large number of references contained in this chapter provides a useful guideline to the relevant literature.

1.1. Models

According to Churchman[1], the system x is the model of the system y (or, x simulates y) if:

i) y is considered the real system

ii) x is considered an approximation of y:

iii) the validity rules of x are not error-free.

Roughly, there are two types of models employed in chemistry[2]: global hard models (GHM's) and local soft models (LSM's).

The GHM's describe the considered chemical systems in terms of fundamental quantities (mass, charge, energy and time). Examples of GHM's are quantum chemical models and kinetic models. The global hard models offer far reaching predictions, but due to mathematical and computational difficulties, they are properly applicable to rather simple systems.

The LSM's apply to restricted classes of chemical systems. In general, the LSM's are semi-empirical models. Typical examples of LSM's are the linear free energy relationships[3a] (LFER) and quantitative structure - activity relations[3b] (QSAR) - in general, the parametric models[5]. The Monte Carlo models belong to the LSM's.

From a mathematical viewpoint, the models may be classified as:

1) analytical models, and

2) (analogical and digital) simulation models.

A computer simulation model is a logical - mathematical representation of a system programmed for solution on a high-speed electronic computer. A fully automatic simulation is a simulation in which
the process is completely automated for computer; if humans play an
integral part in the model processes, the simulation is termed semi-
automatic.

Any computer simulation experiment proceeds through eight
steps[4], namely:

1) problem formulation;

2) formulation of the input and output data;

3) model formulation (algorithm);

4) physical validation of the model;

5) model coding (i.e. computer implementation of the algorithm);

6) program validation;

7) simulation experiments;

8) results validation - this step requires carefully selected statistical methods (as an example, see the work[6]
   of de Maine on random self-avoiding walks on lattices).

Current opinion concerning the utility of simulation lies
between the following two extremes:

"Simulations modelling may alone determine the best design
to minimize the performance degradation..." (Merikallio and Holland[7]),
and

"... Results from any simulation are useful only if the user
actually believes in simulation. An act of faith is required".
(Champbell and Heffner[8]). There is a danger of relying on a simulation
model too heavily but much of the doubt can be alleviated by rigorous validations (i.e. the steps 4,6 and 8).

Further details concerning the computer simulation may be found in refs. 9-13. For a glossary of terms used in gaming and simulation one may consult ref. 14.

1.2. Random Numbers

The numbers, x_1, x_2, ...,x_n in an interval I constitutes, according to von Mises[15], a sequence of random numbers if the following two conditions hold:

i) $(x_i)_{1 \leq i \leq n}$ satisfy some distribution properties;

ii) these distribution properties are invariant under certain selection rules for subsequences extracted from the sequence $(x_i)_{1 \leq i \leq n}$ (for a detailed discussion in this context see ref. 16).

For practical purposes, the random numbers are obtained by means of digital computers according to deterministic (arithmetical) algorithms, i.e. random numbers generators[17,18]. Such numbers will of course not be genuinely random, since they are produced by some deterministic sequence of machine operations. They are normally described as pseudorandom numbers.

The basic random numbers sequence is the sequence of uniform random numbers in the interval $(0,1)$, i.e., $x_i \varepsilon (0,1)$, $i = 1,2,...,n$ and the probability density function is $f(x)=1$ if $x \varepsilon (0,1)$, and $f(x) = 0$ if $x \notin (0,1)$. From a sequence of uniform random numbers one may obtain random numbers with any distribution in any interval I.

The uniform random numbers generators have the form:

$$x_{n+1} = F(x_n); \quad x_1 \quad \text{is given.} \tag{1}$$

The first uniform random numbers generator (termed "mid-square method") was developed by von Neumann. This algorithm, however, is not acceptable because the fraction of smaller values is higher than necessary.

The most used random numbers generators are the congruential generators; namely

i) multiplicative generators:

$$x_{j+1} = Ax_j \text{ (modulo T); } x_1 \text{ is given} \tag{2}$$

ii) mixed generators:

$$x_{j+1} = (Bx_j + C) \text{ (modulo T); } x_1 \text{ is given} \tag{3}$$

The sequence of random numbers generated by these recurrence relations repeats itself at least after T-1 and T steps respectively. We can achieve the maximal cycle length by taking the following values for the constants[139]: T should be given a very large value, the word length of the computer for instance. B is subject to the conditions $T/100 < B < T - T^{1/2}$ and $5 = B \pmod 8$. The constant C should be an odd integer with $C/T \approx 1/2 - 3^{1/2}/6$.

Prior to proceeding to use a uniform random numbers generator, one must make it sure that the generator works well on the desired sequence (x_i), i.e., the sequence is uniformly repartised, and the sequence entries are independent. One may test the uniform repartition computing the values of the moments of the order k = 1,2,3,4:

$$M_k = \frac{1}{N} \sum_{i=1}^{N} x^k, \quad k = 1,2,3,4$$

where N represents the length of the tested sequence. The expected values of M_k are: $M_1 = 0.500$, $M_2 = 0.333$, $M_3 = 0.250$, $M_4 = 0.200$.

To test the independence of the sequence entries one commonly uses the χ^2-statistics:

$$\chi^2 = \frac{n}{N} \sum_{i=1}^{n} (f_i - \frac{N}{n})^2 \tag{5}$$

where N is the length of the random points sequence generated in the interval $\overset{p}{\underset{j=1}{X}} (0,1) = (0,1)^p$, p = 1,2,...,n is the number of equal classes in which $(o,1)^p$ is divided and f_i stands for the number of random points generated in each class i, i = 1,2,...,n. The sequence entries are independent if:

$$\chi^2 < \chi^2_{1-\beta,\ n-1} \tag{5}$$

β is the confidence level, and n-1 represents the degrees of freedom (see any textbook of statistics).

For a chronological and classified bibliography on random numbers generation and testing one may consult ref. 19.

Illustratively, we studied comparatively two multiplicative congruential random numbers generators RANDOM (IMB) and ALEAT (CII), and the mixed congruential generator[20] RAND.

ALEAT and RANDOM are written in FORTRAN, but for IRIS C-50 or FELIX C-256 computers are necessary special procedures (NMAS and MAS) for masking the binary overflow. RAND is written in FELIX C-256 assembler language (the sources of these generators are given within the TEST program).

The results displayed in Table 1 (M_k and χ^2 values) were obtained by means of our TEST program. The flow chart and the listing of the program are given below.

We note that the studied generators work well, i.e., the computed values of the moments are near the expected values, and the generated random points are independent ($\chi^2_{0.95,24} = 36.4$, $\chi^2_{0.99,24} = 43.0$, $\chi^2_{0.95,26} = 38.9$, $\chi^2_{0.99,26} = 45.6$). One may observe small fluctuations of the generators statistical quality, depending on the particular selected sequence and on the sequence length.

Table 1. Results of the experiments performed with three random numbers generators[*)]

Sequence of random numbers	Generator	χ^2 p=1	p=2	p=3	Moments M_1	M_2	M_3	M_4
1-1000	ALEAT	30.650	37.400	36.450	0.512	0.344	0.257	0.205
	RANDOM	27.650	14.900	26.244	0.491	0.325	0.242	0.193
	RAND	20.99	20.550	29.160	0.498	0.331	0.247	0.196
1001-2000	ALEAT	14.750	27.500	23.004	0.488	0.319	0.236	0.187
	RANDOM	31.350	11.200	21.870	0.498	0.336	0.255	0.207
	RAND	24.550	12.400	22.194	0.515	0.351	0.268	0.217
2001-3000	ALEAT	27.750	34.100	45.360	0.503	0.335	0.252	0.203

Table 1 - continued

	RANDOM	17.650	14.700	14.094	0.499	0.333	0.250	0.200
	RAND	19.600	18.100	23.328	0.495	0.330	0.249	0.200
3001-4000	ALEAT	21.650	33.700	23.796	0.493	0.323	0.240	0.190
	RANDOM	30.150	19.200	24.948	0.481	0.318	0.238	0.191
	RAND	24.000	31.700	22.356	0.512	0.348	0.266	0.216
4001-5000	ALEAT	29.200	23.600	27.217	0.506	0.337	0.252	0.201
	RANDOM	27.000	23.900	28.188	0.493	0.330	0.248	0.199
	RAND	36.450	11.800	19.440	0.501	0.333	0.249	0.197
5001-6000	ALEAT	13.050	20.600	28.512	0.499	0.332	0.249	0.201
	RANDOM	23.150	12.600	26.568	0.495	0.328	0.244	0.193
	RAND	10.700	9.600	15.714	0.505	0.338	0.254	0.204
6001-7000	ALEAT	22.450	10.300	14.580	0.492	0.325	0.241	0.191
	RANDOM	27.400	31.200	20.412	0.484	0.322	0.242	0.195
	RAND	22.650	21.700	25.272	0.494	0.330	0.248	0.200
7001-8000	ALEAT	26.800	38.700	26.244	0.505	0.337	0.252	0.200
	RANDOM	19.000	23.100	18.792	0.488	0.321	0.238	0.189
	RAND	24.900	16.800	19.764	0.487	0.320	0.239	0.191
8001-9000	ALEAT	35.200	18.800	40.986	0.507	0.347	0.265	0.215
	RANDOM	41.800	32.200	21.060	0.483	0.316	0.234	0.185
	RAND	22.350	32.600	24.624	0.503	0.337	0.253	0.202
9001-10000	ALEAT	33.350	25.800	21.708	0.496	0.332	0.250	0.201
	RANDOM	29.050	17.200	38.880	0.508	0.340	0.254	0.203
	RAND	33.650	30.700	11.826	0.481	0.316	0.234	0.186
10001 -20000	ALEAT	26.625	17.100	24.435	0.503	0.336	0.252	0.201
	RANDOM	18.730	22.770	14.488	0.505	0.340	0.257	0.207
	RAND	24.395	19.470	24.759	0.499	0.332	0.249	0.199
1-100000	ALEAT	31.939	18.300	33.996	0.502	0.335	0.252	0.201
	RANDOM	12.281	9.003	23.568	0.501	0.335	0.251	0.201
	RAND	22.604	15.819	12.209	0.500	0.333	0.250	0.200

(*) For p=1, n=25; p=2, n=25; p=3, n=27.

8

FIGURE 1. FLOWCHART OF TEST PROGRAM

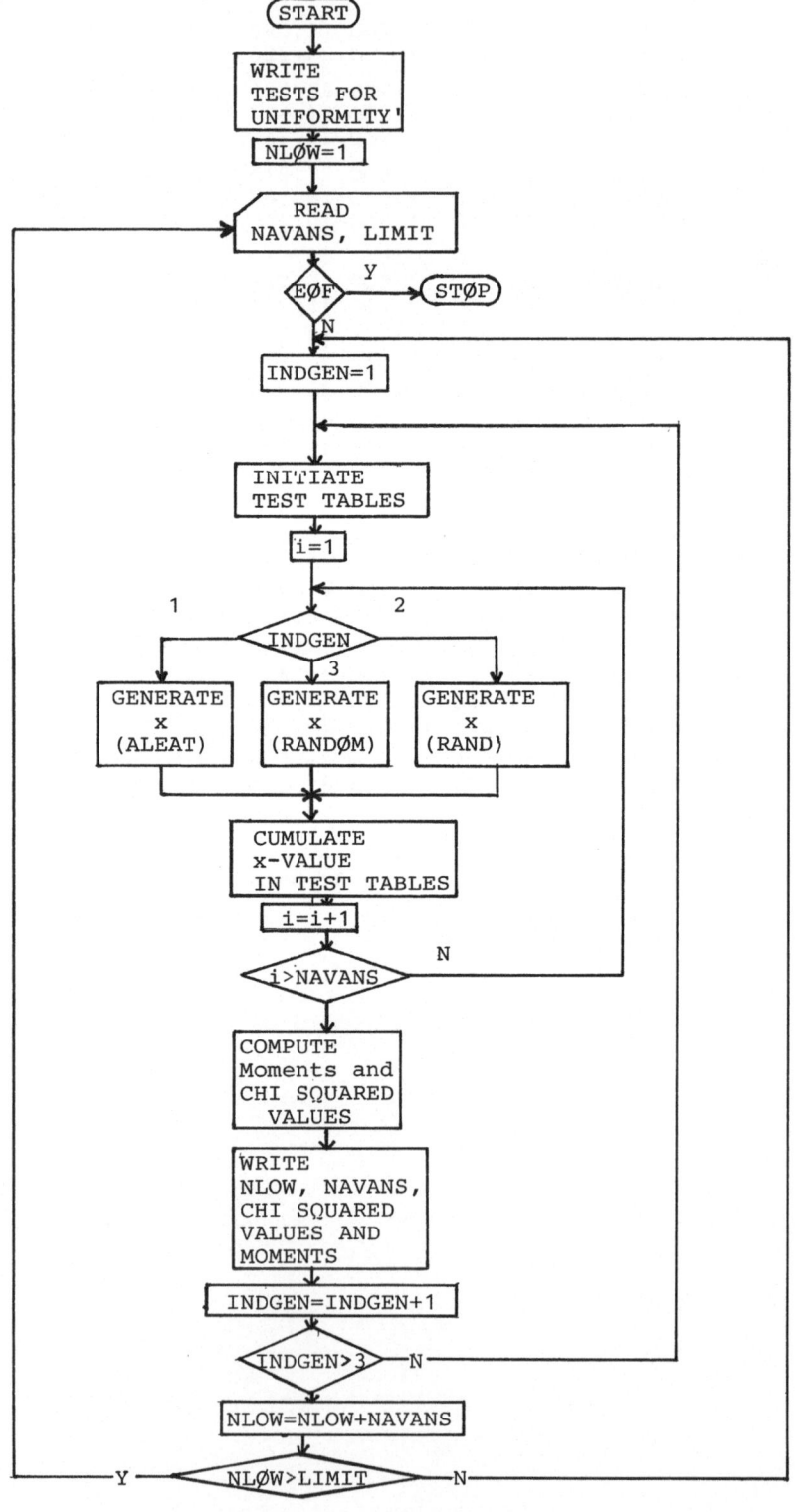

THE TEST PROGRAM
----

1. FORTRAN ROUTINES

```
        DIMENSION FRECV(30), COMUNIC(30), SQUARE(5,5), CUB(3,3,3),
     *           IXI(2), ICUB(3), T(30,3), NI(3), CHI(3), ID(3), RN(3),
     *           RMOMENT(4), GN(2,3), GCHI(3,3), GMOM(4,3)
        DATA    GN / 'IIALE','AT II','IIRAN','DOMII','IIRAN','D  II'/
        DATA    IA, NI / 65539, 25, 5, 3/
        DATA    NI1, NI2, NI3 / 25, 5, 3/
C
        DO 98 L = 1, 3
98      ID(L) = NI(L) ** L
        WRITE (108, 67 )
        WRITE (108, 400)
        WRITE (108, 401)
        WRITE (108, 400)
        NLOW = 1
888     READ (105, 96, END = 1113) NAVANS, LIMIT
96      FORMAT ( 2I8 )
889     CONTINUE
        DO 1 INDGEN = 1, 3
        DO 40 I = 1, NI1
40      FRECV(I) = 0.0
        DO 50 I = 1, NI2
        DO 50 J = 1, NI2
50      SQUARE(I,J) = 0.0
        DO 51 I = 1, NI3
        DO 51 J =1, NI3
        DO 51 K = 1, NI3
51      CUB(I,J,K) = 0.0
        ISQ = 0
        IC  = 0
        DO 661 I = 1, 4
661     RMOMENT(I) = 0.0
        DO 100 I = 1, 3
100     RN(I) = FLOAT(NAVANS) / FLOAT(I)
C
C       TEST  SEQUENCE
C
        DO 3 I3 = 1, NAVANS
        GOTO (1109, 1110, 1111), INDGEN
1109    CONTINUE
        CALL MAS
        CALL ALEAT (IA, IB, X)
        IA = IB
        CALL NMAS
        GOTO 1112
```

```
1110   CONTINUE
       CALL MAS
       CALL RANDOM (X)
       CALL NMAS
       GOTO 1112
1111   CONTINUE
       CALL RAND (X)
1112   CONTINUE
       DO 662 I = 1, 4
662    RMOMENT(I) = RMOMENT(I) + X ** I
       ISQ = 1 + MOD(ISQ,2)
       IXI(ISQ) = 1 + INT(X * NI2)
       IF (ISQ .EQ. 2) SQUARE(IXI(1), IXI(2)) = SQUARE(IXI(1), IXI(2))
      *                 + 1.0
       IC = 1 + MOD(IC, 3)
       ICUB(IC) = 1 + INT(X * NI3)
       IF(IC .EQ. 3) CUB(ICUB(1), ICUB(2), ICUB(3)) =
      *               CUB(ICUB(1), ICUB(2), ICUB(3)) + 1.0
       IX = 1 + INT(X * NI1)
3      FRECV(IX) = FRECV(IX) + 1.0
C
C      EFFECTIVE TESTS AND LISTING
C
       DO 70 I = 1, NI1
70     T(I,1) = FRECV(I)
       DO 71 I = 1, NI2
       DO 71 J =1, NI2
       K = (I-1) * NI(2) + J
71     T(K, 2) = SQUARE(I, J)
       DO 72 I = 1, NI3
       DO 72 J = 1, NI3
       DO 72 K = 1, NI3
       L = (I-1) * NI(3) ** 2 + (J-1) * NI(3) + K
72     T(L, 3) = CUB(I, J, K)
C
       DO 88 J1 = 1, 3
       LID = ID(J1)
       DO 89 J2 = 1, LID
89     COMUNIC(J2) = T(J2, J1)
       CALL CHITEST    (COMUNIC, RN(J1), ID(J1), CHI(J1))
88     CONTINUE
       DO 60  I = 1, 3
60     GCHI(I, INDGEN) = CHI(I)
       DO 61 I = 1, 4
61     GMOM(I, INDGEN) = RMOMENT(I) / RN(1)
1      CONTINUE
       NLW = NLOW - 1
       WRITE (108, 402) NLW, NAVANS
      *,              ( (GN(I, J), I = 1, 2)
      *,                (GCHI(I, J),I = 1, 3)
      *,                (GMOM(I, J),I = 1, 4), J = 1, 3)
       NLOW = NLOW + NAVANS
       IF (NLOW .GT. LIMIT) GOTO 888
       GOTO 889
```

```
1113    CONTINUE
        WRITE (108, 400)
        WRITE (108, 1114)
        STOP
67      FORMAT('1'/26X,'TESTS FOR UNIFORMITY'/26X,20('*')//)
400     FORMAT(' ',8X,71('='))
401     FORMAT(' ',8X,
       *'||REJEC.||ACCEP.|| GEN. || CHI SQUARED VALUES ||',
       *'       MOMENT VALUES          ||'/9X,
       *'||RANDOM||RANDOM||           ||--------------------||',
       *'--------------------------||'/9X,
       *'||NUMBER||NUMBER|| NAME ||   1   ||   2   ||   3   ||',
       *'  M1   ||  M2   ||  M3   ||  M4   ||'/9X,
       *'||      ||      ||       || DIM. || DIM. || DIM. ||',
       *'  0.50 || 0.(3)|| 0.25 || 0.20 ||')
402     FORMAT(' ',8X,2('||',I6),2A4,7(F6.3,'||')/
       *(9X,2('||        '),2A4,7(F6.3,'||')))
1114    FORMAT(' '/26X,'END OF TESTING'/26X,14('*'))
        END


        SUBROUTINE ALEAT(I,J,Z)
        J = I * 65539
        IF (J .LT. 0) J = J +2147483647 + 1
        Z = J
        Z = Z * 0.4656613E-9
        RETURN
        END


        SUBROUTINE RANDOM(Z)
        DATA I /1/
        INTEGER A, X
        IF (I .EQ. 0) GOTO 1
        I = 0
        M = 2 ** 20
        FM = M
        X = 566387
        A = 2 ** 10 +3
1       X = MOD(A * X, M)
        FX = X
        Z = FX/FM
        RETURN
        END


        SUBROUTINE CHITEST (A, O, K, C)
        DIMENSION A(K)
        XK = FLOAT (K)
        S = 0.0
        DO 1 I = 1, K
1       S = S + (A(I) - O / XK) ** 2
        C = S * XK / O
        RETURN
        END
```

2. ASSEMBLER ROUTINES

```
        CSECT
        DEF         MAS,NMAS,RAND,RANDLIST
A       DATA,4,4    65529
X       DATA,4,4    0
N       DATA,4,4    0
M       DATA,4,4    X'052E0000'
B       DATA,8,8    3
T       EQU         X
CARACT  DATA        62
RS8     DATA,4,4    0
MASK    DATA,4,4    X'50400000'
*
*       BICONGRUENTIAL RANDOM NUMBER GENERATOR
*       ACCESS BY: CALL RAND(X)
*
RAND    LD4,14      %
        LD4I,3      4
        MVSR,2      X
RAND1   LD4,0       A
        MP4,0       M
        SRL8,0      1
        LD4I,0      0              (KM*X)MOD 2**32
        AD8,0       B              ((KM*X)MOD 2**32 + KA)MOD 2**32
        LD4I,0      64
        ST4,1       M
        LD4,1       4
        BCF,4       POZ
        SRL8,0      8
STX     ST4,1       *X
        BRU         *32
POZ     NF4,1       CARACT
        BRU         STX
*
RANDLIST LD4,14     %
        LD4I,3      8
        MVSR,2      T
        ST4,8       RS8
        LDC4,3      *N
RAND2   BAL,8       RAND1
        IC4,4       T
        IC4,1       12
        BZ          *RS8
        BRU         RAND2
*
MAS     LD4,14      %
        LDTM,13     MASK
        BRU         *32
*
NMAS    LD4,14      %
        LDTM,13     MASK+1
        BRU         *32
        END
```

Often random numbers are required with distributions other than the rectangular ones discussed so far. For the more commonly used distributions rather efficient transformation methods have been developed. They allow the calculation of the respective random numbers from uniformely distributed ones $x \in (0,1)$. Here we present only a few methods which are important for our purpose.

1.) Rectangular distribution

$$f(x) = \begin{cases} 1/(b-a) & \text{if } x \in (a,b) \\ 0 & \text{if } x \notin (a,b) \end{cases} \tag{6}$$

$$y_i = (b-a)x_i + a \tag{7}$$

2.) Poisson distribution

$$f(k) = \frac{a^k}{k!} e^{-a} \tag{8}$$

This distribution can be generated by a method illustrated in Fig. 2

3.) Normal distribution

$$f(x) = (2/\pi)^{1/2} \exp(-x^2/2)$$

One of the simplest but nevertheless effective methods is based on the central-limit-theorem, according to which the random variable

$$Z = \sum_{k=1}^{12} x_K - 6 \tag{9}$$

is asymptotically normally distributed.

4.) Sine and cosine of uniformly distributed angles ϕ (Fig. 3).

5.) Generation of unit vectors uniformly distributed in 3-dimensional space (Fig. 4).

1.3. Computer Monte Carlo Models

Monte Carlo is a numerical method of solving stochastic models without determination of the analytical representations of the system.

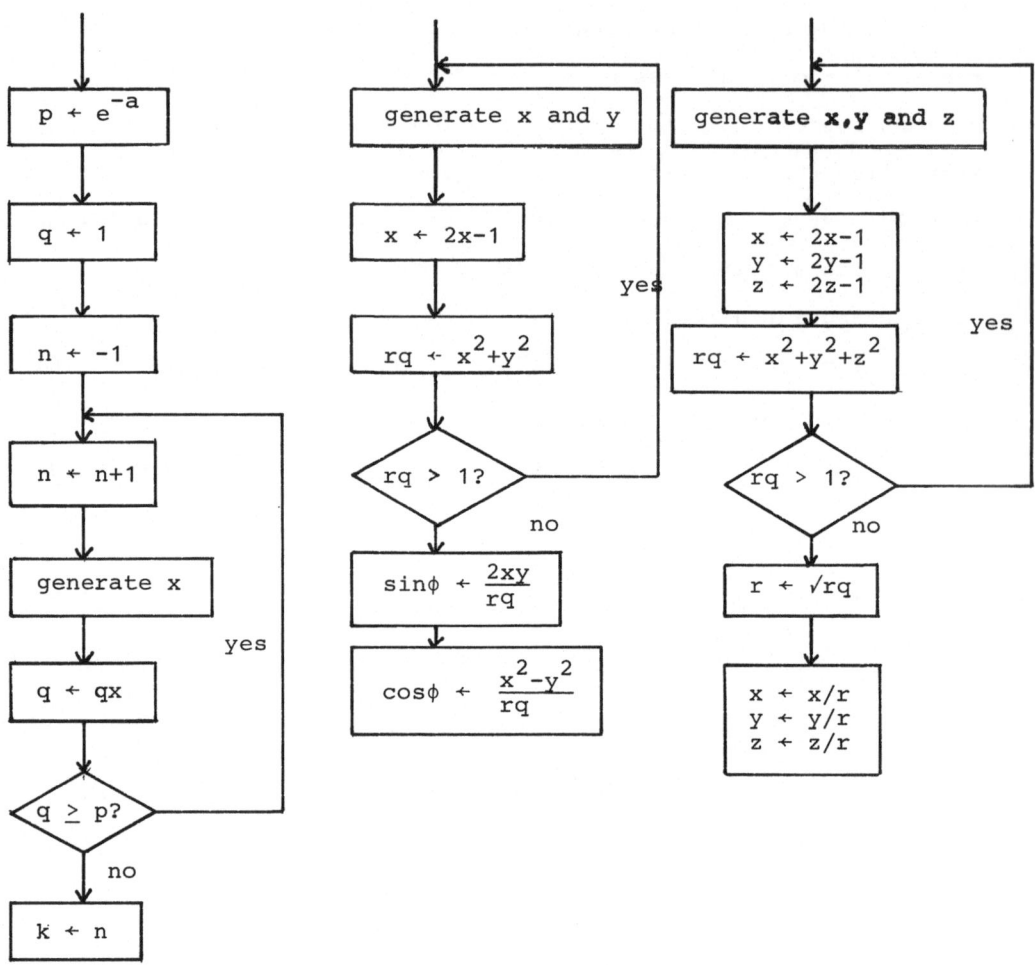

Figure 2   Figure 3   Figure 4

The heart of Monte Carlo method is the usage of random numbers. The birth of this method is assumed to have taken place in 1949 with publication of the paper by Metropolis and Ulam[21], although it was known earlier[22] to von Neumann, Metropolis, Ulam, Kahn and their collaborators working on U.S. Defense projects at Los Alamos Scientific Laboratory.

The computer Monte Carlo models are computer models based upon the Monte Carlo method.

The references 23-34, 37, 38 explore the whole range of the Monte Carlo method and its applications. For a history of the random-walk problem one may consult ref. 35. Ref. 36 contains a selected bibliography of Monte Carlo applications in Polymer Science.

A possible formulation of a Monte Carlo model may consist in the following steps:

    i) identify the states E_1, E_2, ..., E_n of the considered real system;

    ii) compute the transition probabilities p_1, p_2, ..., p_n. The system exists in the state E_j with the probability p_j.

    iii) generate the random number $x \in (0,1)$. If the inequalities (10) hold

$$\sum_{j=1}^{k} p_j < x \leqslant \sum_{j=1}^{k+1} p_j \tag{10}$$

the state E_k occurs. This step is repeated as many times as desired.

There may exist any type of restrictions in connection with inequality (10). It is very easy to account for them in the framework of the Monte Carlo models.

It is clear that the Monte Carlo model does not represent the behaviour of the real system directly. The relation between the Monte Carlo model and the simulated real system is[11]: by means of a

suitable Monte Carlo model one constructs a state history, in chronological order, of the real system dealt with. The state history consists of a succession of state descriptions corresponding to the
states of the system. The chronological succession of the states
of the model is regarded as the state history of the real system.
The results of the simulation are validated comparing a partial state
history produced by the model with the same partial state history
produced by the real system.

1.3.1. The accuracy of the Monte Carlo method

Let us suppose that we compute the observable c, by means of
a correctly defined Monte Carlo model. The expected value of c is
$E(c)$. In a simulation run one obtains for c the value c_I. Using
another random numbers sequence and recomputing the value of c, one
gets c_J, $c_I \neq c_J$. Through N simulations runs, the average value of
c is $\bar{c}$:

$$\bar{c} = \frac{1}{N} \sum_{I=1}^{N} c_I. \qquad (11)$$

From the law of the large numbers one gets:

$$\sigma^2(\bar{c}) = \sigma^2(c)/N = E \{[c - E(c)]^2\}. \qquad (12)$$

Passing to the limit in the relation (12), one obtains

$$\lim_{N \to \infty} \sigma^2(\bar{c}) = \lim_{N \to \infty} \sigma^2(c)/N = 0 = \lim_{N \to \infty} E\{[\bar{c} - E(c)]^2\}. \qquad (13)$$

Relation (13) states that $\bar{c}$ becomes concentrated about $E(c)$ as N
increases (σ^2 stands for variance). Thus, the precision of the Monte
Carlo calculations depends on the value of N.

In order to compute what N should be, required to ascertain
a given accuracy for $\bar{c}$, one uses the Tchebycheff inequality:

$$P\left[|\bar{c} - E(c)| > b \right] < \frac{\sigma^2(c)}{b^2 N} \qquad (14)$$

where $\sigma^2(c)$ is estimated by the sample variance

$$s^2 = \frac{1}{N-1} \sum_{I=1}^{N} (c_I - \bar{c})^2 .$$

An equivalent form of inequality (14) is

$$P\left[\bar{c}-b < E(c) < \bar{c}+b\right] > 1 - \frac{\sigma^2(c)}{b^2 N} . \tag{15}$$

The relation (15) states that $E(c)$ belongs to the interval
$(\bar{c} - b, \bar{c} + b)$ with a probability greater than $1 - \sigma^2(c)/(b^2 N)$.
The relations (14) or (15) are useful to compute the accuracy of the
Monte Carlo results, or to determine the value of N for which a
given precision is obtained.

The above statments are illustrated by the examples below.

1. The binary irreversible copolymerization experiments performed with our MEMORY-3 program (see Chapter 2) furnished[39] the
following dependence of the molar percent of the mer M_1 on the polymerization degree:

%M_1	10	50	100	200	300	400	500	600
polyme- rization degree	90.oo	60.oo	57.00	57.00	56.67	56.00	56.00	56.17
	700	800	900	1000				
	56.17	56.25	56.33	56.31				

Thus, one assures an acceptable stability considering a polymerization degree $\geqslant$ 500.

2. The Monte Carlo results are very sensitive to the statistical quality of the used sequence of random numbers. In order to
illustrate this statement we used 100,000 random numbers to simulate
100 macromolecules with polymerization degree 1000 (i.e., the sequences 6000-7000, 7000-8000 etc. were used to simulate the polymer
macromolecules). The numerical data collected[39] in Table 2 (molar
percent of mer M_1 in macromolecule) refer to the penultimate effect

copolymerization of benzylidenecyanoacetic ester (M_1) and styrene (M_2); with r_1=0.325, $r_1' = 1.330$, $r_2 = r_2' = 0.000$, molar fraction of M_1 in feed is 0.6, according to ref. 40.

Table 2. The dependence of the Monte Carlo results on the statistical quality of the used random numbers sequence.[a]

No.	%M_1	No.	%M_1	No.	%M_1	No.	%M_1
1.	66.7	26.	66.1	51.	65.2	76.	65.2
2.	67.1	27.	66.7	52.	66.2	77.	66.7
3.	66.2	28.	66.3	53.	66.8	78.	66.6
4.	65.9	29.	65.2	54.	67.0	79.	66.9
5.	67.3	30.	67.6	55.	67.8	80.	66.8
6.	67.6	31.	65.7	56.	67.6	81.	66.9
7.	67.7	32.	66.9	57.	66.3	82.	66.2
8.	65.7	33.	66.6	58.	67.0	83.	67.3
9.	67.3	34.	66.2	59.	67.1	84.	67.6
10.	65.3	35.	67.0	60.	67.7	85.	66.4
11.	66.6	36.	65.9	61.	64.6	86.	66.5
12.	66.7	37.	67.0	62.	67.3	87.	66.8
13.	66.8	38.	65.9	63.	67.7	88.	67.3
14.	67.3	39.	65.4	64.	67.3	89.	65.8
15.	65.9	40.	65.5	65.	66.4	90.	66.3
16.	66.6	41.	66.3	66.	65.3	91.	66.6
17.	66.5	42.	66.7	67.	66.7	92.	65.4
18.	66.1	43.	66.1	68.	66.6	93.	66.7
19.	67.5	44.	67.1	69.	66.9	94.	66.4
20.	66.0	45.	66.2	70.	67.2	95.	66.3
21.	65.9	46.	67.1	71.	66.1	96.	66.9
22.	66.3	47.	67.2	72.	66.1	97.	66.1
23.	68.4	48.	66.1	73.	66.9	98.	66.5
24.	67.7	49.	66.2	74.	67.3	99.	67.0
25.	67.4	50.	65.9	75.	66.1	100.	67.1

a) The total computing time is 5'1" on the FELIX C-256 computer, using our MEMORY-5 program.

The %M$_1$ values displayed in Table 2 have the standard deviation

$$x = \left[\frac{1}{99} \sum_{i=1}^{100} (\%M_{1,i} - \%\overline{M}_1)^2 \right]^{1/2} = 0.693$$

where

$$\%\overline{M}_1 = \frac{1}{100} \sum_{i=1}^{100} \%M_{1,i} = 66.574$$

Choosing a satisfying (from statistical viewpoint) random numbers sequence one may credit the results obtained in one run. In this way we obtain the important advantage of visualizing the features of the macromolecule (i.e., "... to visualize a polymer chain in all its structural chaos or order" - Price [41]).

It seems that Monte Carlo experiments are a valuable tool for many branches of physics and chemistry, and "... since their application is relatively simple they will become a standard method of scientific research in the near future" (Binder [32]).

1.3.2. Variance reduction methods

As has been shown above the accuracy of the Monte Carlo method is of order $N^{-1/2}$. This means that a tenfold improvement in precision requires a hundredfold increase in the number of samples. To save time and costs attention has been concentrated upon reducing the variance of the sampling process. Several reduction methods have been developed, such as correlated sampling, stratified sampling, importance sampling, antithetic variables. None of these methods can be applied without care, since their effectivity depends on the respective problem. Problems in statistical mechanics are most frequently treated by the method of Metropolis et al.[84]. In the following we give an outline of this method:
The canonical ensemble average of a function of the coordinates of

a system of N particles is defined as

$$\langle f(\{\vec{r}\}) \rangle = \int (f(\{\vec{r}\}) \exp\left[-U(\{\vec{r}\})/(kT)\right] d\{\vec{r}\}/$$

$$\int \exp\left[-U(\{\vec{r}\})/(kT)\right] d\{\vec{r}\} . \qquad (16)$$

A Monte Carlo estimate for $\langle f \rangle$ can in principle be obtained from

$$\langle f \rangle \approx \sum_{i=1}^{m} f(i)\exp\left[-U(i)/(kT)\right]/ \sum_{i=1}^{m}\exp\left[-U(i)/(kT)\right] . \qquad (17)$$

This means m configurations drawn from the parent population of M possible configurations have to be generated, each of them consisting of N position vectors uniformly distributed within the given volume. The Boltzmann factors and sums have to be evaluated, and we obtain the result. This plain sampling method looks indeed simple, but it does not work in practice when using realistic potentials, since a large proportion of the configurations generated are highly improbable because of intermolecular repulsions. Therefore they contribute very little to the sums. To overcome this difficulty some form of impor- tance sampling is used whereby configurations are selected according to the probability

$$\pi_i = \exp\left[-U(i)/(kT)\right]/ \sum_{j=1}^{M} \exp\left[-U(j)/(kT)\right] . \qquad (18)$$

In this case equation (17) has to be modified by the weighting factor π_i^{-1}

$$\langle f \rangle \approx \sum_{i=1}^{m} f(i)\ \pi_i^{-1}\exp\left[-U(i)/(kT)\right]/ \sum_{i=1}^{m} \pi_i^{-1}\exp\left[-U(i)/(kT)\right] = \frac{1}{m} \sum_{i=1}^{m} f(i)$$

$$(19)$$

The sample is constructed here via a Markov process. In contrast
to the plain method the samples are no longer independent, but are
considered as members of a Markov chain. The transition probabilities
for passing from a configuration to the succeeding one must have
certain properties to assure that the average of f over all states
of the chain converges to <f>. In other words, the elements p_{ij} of
the transition matrix must satisfy the condition

$$\pi_j = \lim_{n \to \infty} p_{ij}^{(n)} \text{ (for every state i and j: ergodicity condition),}$$

(20)

$p_{ij}^{(n)}$ is the n-step transition probability defined by the
recurrence relation

$$p_{ij}^{(n)} = \sum_{r=1}^{M} p_{ir} \, p_{rj}^{(n-1)} .$$

A consequence of the ergodicity condition is the so-called steady-
state condition (see the Appendix at the end of this chapter).

$$\pi_j = \sum_{i=1}^{M} \pi_i \, p_{ij} .$$

(21)

This equation can be satisfied by the condition of microscopic
reversibility $\pi_j p_{ji} = \pi_i p_{ij}$,
from which follows

$$\frac{p_{ij}}{p_{ji}} = \frac{\pi_j}{\pi_i} = \exp\{-[U(j)-U(i)]/(kT)\} .$$

(22)

This equation does not uniquely specify the transition matrix. A
commonly used choice is

$$p_{ij} = \begin{cases} 1 & \text{if } \pi_j \geq \pi_i \text{ or } U(j) \leq U(i) \\ \dfrac{\pi_j}{\pi_i} & \text{if } \pi_j < \pi_i \end{cases} .$$

(23)

A realization of this method is described in one of the next sections.

1.4. Monte Carlo applications in Polymer Science

We begin this section with a brief review of Monte Carlo applications in Chemistry.

A. Regression analysis is the basic mathematical tool[44] in searching for structure-reactivity, or chemical structure - biological activity relationships (LFER and QSAR, respectively). In this technique a number of variables (physicochemical parameters, molecular orbital, or structural parameters - denoted collectively by x_{ij}) are tested for correlation with a suitable measure of reactivity (i.e., rate constants k, or equilibrium constants, K) or biological activity (LD_{50}, ID_{50} etc.) denoted by y_i).

In general, a linear model is assumed:

$$\hat{y}_i = a + \sum_j b_j x_{ij} ; \qquad \begin{array}{l} i = 1,2,\ldots, n \\ j = 1,2,\ldots, m \end{array} \qquad (24)$$

where $\hat{y}_i$'s estimate the measured y_i values. The constants a and b_j are computed by the least squares method, using the condition:

$$\Delta y = \sum_{i=1}^{n} (y_i - \hat{y}_i)^2 = min! \qquad (25)$$

Topliss and Costelle[42] studied the role of chance in the observed correlations (24). Toward this end, they used the Monte Carlo method in the following manner: one generated at random the values y_i of the dependent variable and corresponding sets of independent variables, x_{ij}, and correlations (24) were computed. These authors concluded that regressions possessing correlation coefficients $r^2 < 0.40$ are obtained by chance, and that the minimum number, n, of observations (i.e., y_i values) required to test m independent variables (i.e., x_{ij} parameters) is:

n	30	50	65	85
m	5	10	20	30

The next paper of the series[43] investigates the risk of arriving at fortuitous correlations when too many variables are screened relative to the number of available observations.

B. Hydrophobic interactions are responsible for the stability of particular conformations of large molecules in aqueous solutions, for the stability of micelles and biological membranes, and play a decisive role in biological receptor/bioactive molecule interactions. Several Monte Carlo studies have recently appeared on the topic of hydophobic interactions[45-50]. In general, one considers a system composed[50] of a number of water molecules and a pair of spherical nonpolar chemical species. The water molecules interact with each other and with the nonpolar species through a given potential (in general, Lennard-Jones type). Then, one generates, by means of the Monte Carlo technique, a large number of configurations of the system and computes the pair correlation function, $g(r)$, for the apolar species. These simulations pointed out[50] that there are two relatively stable conformations for the apolar species: one in which each member of the pair sits in its own water cage with one water molecule between them, and the second one in which no water molecule sits between the spheres,

$$A(H_2O)A \quad \rightleftarrows \quad A \ldots A$$

conformation I   conformation II

C. For the study of interactions in biological systems (i.e., enzyme systems, nucleic acid/large ligand interactions etc.) the Monte Carlo approach seems to be an extremely powerful one, the primary virtue of this approach being its outstanding flexibility. Monte Carlo studies on topics as interacting enzyme systems at steady state[51-53] kinetics of the cooperative helixcoil transitions[54] kinetics of nucleic acid/large ligands interactions[55] etc. have appeared. In general, the studied systems are represented by lattices (one - dimensional lattice in ref. 55, two - dimensional lattice in ref. 51-53), and a Monte Carlo model generates the equilibrium lattice configurations.

D. Motoc et al.[56-59] (for an up-to date review, see ref. 60) used the Monte Carlo method to treat quantitively the steric effects in chemical reactions[56-58] (OVA method) and in biological systems[59] (MCD method). We proved[58] that in reaction series the steric repulsion energy, δE is given by an equation of the form:

$$\delta E = \text{const. } OV(r)$$

where OV is the volume $(\mathring{A}^3)$ of the van der Waals envelopes of the partners which overlap in the transition state and r specifies the position of the considered transition on the reaction coordinate. The OV values are computed by means of the Monte Carlo method[56,57]. Used in concrete applications, the OVA method proved its correlational ability. Using the Monte Carlo method, MCD computes the sum of the nonoverlapping van der Waals volumes of two superimposed molecules. MCD has been utilized as a steric parameter in QSAR[59].

E. Monte Carlo computer simulations of the structure of the liquid[61-64, 68-71] water or its imperfect vapors[65] have been reported. These simulations offered qualitative conclusions concerning some structural properties of water. Other Monte Carlo studies are concerned with the distribution of water molecules around an ion pair (for Li^+F^- pair, see ref. 66,67), the properties of HF (ref. 72,73), Na-K alloy[74], Rb(ref. 75), Na, Cs(ref. 76), Al (ref. 77), Li (ref. 78), liquid potassium cyanide[79], molten caesium halides[80], molten potassium chloride[81] etc.

The remaining of the chapter is devoted to review the Monte Carlo applications in Polymer Science.

1.4.1. Solvent around biomacromolecules

Monte Carlo studies of water structure around biomacromolecules were performed by Hagler and Moult[82]. The heart of their method is to generate enough states of the water in the system to allow the calculation of the desired macroscopic property X by the weighted averaging over the states (Metropolis method[84]):

$$<X> = \left[\sum_i X \exp(-E_i/kT) \right] / \left[\sum_i \exp(-E_i/kT) \right]. \qquad (26)$$

Using the relation (26), one may compute structural (i.e., mean positions and temperature factors of the ordered water molecules, the probability of finding a water molecule at a given position), or energetic (i.e., the average energy and energy distributions of water in different environments) properties of the considered system. Hagler and Moult used an algorithm that is a realization of the Markov chain described in Section 1.3.2.

1. Select a starting configuration in which the water molecules are arbitrarily packed around the solute biomacromolecule to give a density equal to that of bulk water.

2. Select at random individual water molecules and translate at random with 0.25 Å.

3. Select at random an axis of rotation (x,y or z) and rotate randomly the molecule by 10°.

4. Compute the change in energy of the system. If the change is favourable, the generated configuration of the system is accepted. If the change in energy is unfavourable ($\Delta E>0$), the Boltzmann factor $\exp(-\Delta E/kT)$ is compared with the random number $x \epsilon (0,1)$. If

$\exp(-\Delta E/kT) > x$ the configuration is accepted, otherwise it is rejected, and the old configuration is counted again in the configurational average (17).

    5. Go to the step (2), as many times as necessary.

    6. Compute the statistical average <X> of the property X.

An important problem in these computations is the choice of potentials used in calculating energies.

The results obtained in the case of the hexapeptide cyclo-(L-Ala-L-Pro-D-Phe)$_2$ crystal are shown in Table 3.

Table 3. Cyclic hexapeptide crystal: simulated and experimental ordered water positions[x)]

Water molecule	Rowlinson water model		Stillinger water model		X-ray
	Deviation from X-ray position	R.m.s. movement (simulation)	Deviation from X-ray position	R.m.s. movement (simulation)	R.m.s. vibration
On twofold axis	0.6	0.72	0.8	0.71	0.49
In a general position	0.5	0.52	0.2	0.58	0.45
	0.5	0.50	0.4	0.50	0.45
	0.4	0.56	0.6	0.69	0.45
	0.4	0.52	0.5	0.54	0.45

x) All quantities are given in Å; averages are based on about 140,000 configurations after convergence.

R.m.s. stands for the average root mean square deviation of molecules from their mean position. Simulation results reproduce the experimental data with accuracies of about 0.5 Å.

The results obtained in this study (i.e., there exist water channels of low energy water molecule which "glue" together protein molecules 6-7 Å apart from each other, water molecules more

than 4.2 Å from any protein atom are similar to bulk water etc.)
argue the statement[83] that "naive models of water structure around
biological molecules have not been particularly useful, and that
computer descriptions of water behaviour are necessary in order
to model the irreducible complexity of such systems".

1.4.2. Self-replicating macromolecules

The interest in self-replicating macromolecules is justified
due to the biological importance of these systems (i.e., DNA and protein synthesis) and the technological importance of synthetic macromolecules which are capable of selectively replicating copolymers.
Frisch, Bishop and Roth[85-87] have investigated using Monte Carlo
computer models the chemical mechanisms which are operative in this
complex kinetic scheme.

The involved model considers an infinite porous catalyst
viewed as a one-dimensional lattice with N locations containing
monomers (denoted by H-A-OH and H-B-OH) and polymers (i.e., the
homopolymers H-A-A-...-A-OH, H-B-B-...-B-OH and the random copolymer
H-A-A-B-A-B-B-...-A-OH). A schematic view of the model is given in
Figure 5.

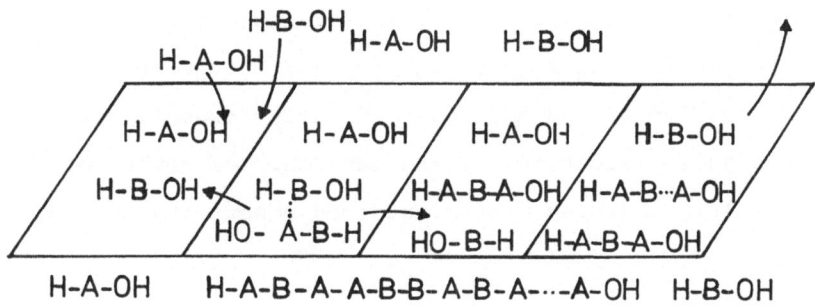

Figure 5. Schematic view of Frisch et al. model (the arrows
indicate the possible movements).

The model considers the following processes[86]:

1. growth:

$$H-A-OH + H-B-B-B-OH \xrightarrow{k_g} H-A-B-B-B-OH + H_2O, \text{ or}$$

$$H-B-B-B-A-OH + H_2O$$

2. absorption:

$$H-A-A-OH + H-B-B-OH \xrightarrow{k_a} H-A-A-B-B-OH + H_2O, \text{ or}$$

$$H-B-B-A-A-OH + H_2O$$

3. complementation (binding):

```
        H-B-OH                     H-B-OH
          ↺          k_b              ⋮
H-B- B-A-A-OH ───────→ H-B-B-A-A-B-OH
```

```
        H-A-OH
          ↙
    H-B-OH
      ⋮
H-B-B-A-A-B-OH ─//→
```

4. replication:

```
H-B-B-A-B-A-OH              H-B-B-A-B-A-OH
 ⋮ ⋮ ⋮ ⋮ ⋮      k_r            +
HO-B-A-B-A-B-H ───────→ HO-B-A-B-A-B-OH
```

The monomers may be absorbed on the catalyst through their unreacted functional groups. Polymer chains are mobile: the higher is the polymerization degree, the greater is the tendency to desorb from the catalyst. Randomly selected chains execute a random walk between adjacent locations of the lattice. The number of monomers (N_A and N_B) in the system is maintained constant (i.e., $N_A + N_B$ = constant).

The following rules allow for evolution in a competitive millieu:

1. When the chain is moved from a location into another (non-empty) location of the lattice it interacts with either mono-

mer or another chain. The probability of each interaction is equal
with the fraction of the species present (the moved chain is excluded
from this fraction).

2. If the moved chain with polymerization degree l inter-
acts with a monomer, the uniform random integer $y \epsilon$ {1,2,...,l+1} is
selected. If $1 \leqslant y \leqslant l$ the monomer will attempt to bind with the
chain at the position y (complementation); if the complementary
position is occupied or if the attempt is unsuccessful, the move
is over. If y = l + 1, the chain will grow by adding one monomeric
unit. All probabilistic events are realized if a uniform random
number $x \epsilon (0,1)$ is less or equal to the probability of success.

3. The partially completed complement is maintained till
all l positions of the chain are filled. When complement is complete
it separates from the template chain and acts as a template.

4. Chains are allowed to grow up to the length of l = 30 at
which moment only binding is allowed. Chains of length $l \geqslant 21$ may
randomly desorb from the catalyst (uniform random numbers $x \epsilon (0,1)$
modelate these events).

5. When two chains interact, the probability P of successful
interactions is:

$$P = \frac{1}{2} A (M_1^{-1/2} + M_2^{-1/2})$$

where M_1 and M_2 stand for the mass of the chains, and A is a system
parameter.

All probabilities involved in the model are considered as
parameters of the model. The dynamics of the system have been followed for 3×10^6 cycles. Typical results are presented in ref. 87.

1.4.3. Molecular weight distribution

Let us consider (according to Lowry[88]) a homopolymer chain consisting of n mers, $\sim M_n^*$ ($*$: . , + or -), which can either propagate (with the probability p), depropagate (with the probability d), or terminate (with the probability t; p+d+t=1):

$$\sim M_n^* + M \rightleftharpoons \sim M_{n+1}^* \qquad \text{("live" polymer)}$$

$$\sim M_n^* \longrightarrow P_n \qquad \text{("dead" polymer).} \qquad (27)$$

The flow chart of the program[88] which implements the Monte Carlo model for the system (27) is shown in Figure 6. The model conforms to the general Monte Carlo model discussed in the subsection 1.3.

The simulation was run for 1000 macromolecules. Runs were made for t=0.001, 0.003, 0.01, 0.03 and 0.1, and for each t the value of p/(p+d) was from 0.1 to 1.0, at intervals of 0.1.

The "synthesized" macromolecules were analysed by computing the polydispersity index [89,90] Δ:

$$\Delta = \left[(<DP^2> - <DP>^2) / <DP>^2 \right]^{1/2}. \qquad (28)$$

DP is the degree of polymerization. The values of Δ computed by means of the Monte Carlo model were compared with Δ values calculated with analytical relations. Typical results are shown in Figures 7 a and 7 b.

1.4.4. Reactivity in polymer analogous reactions

The reactivity of a mer unit in most polymer analogous reactions depends upon the nearest-neighbouring units. The change in reactivity may be positive or negative. In the first case one speaks of auto-cooperative (auto-catalytic) reactions, in the second of

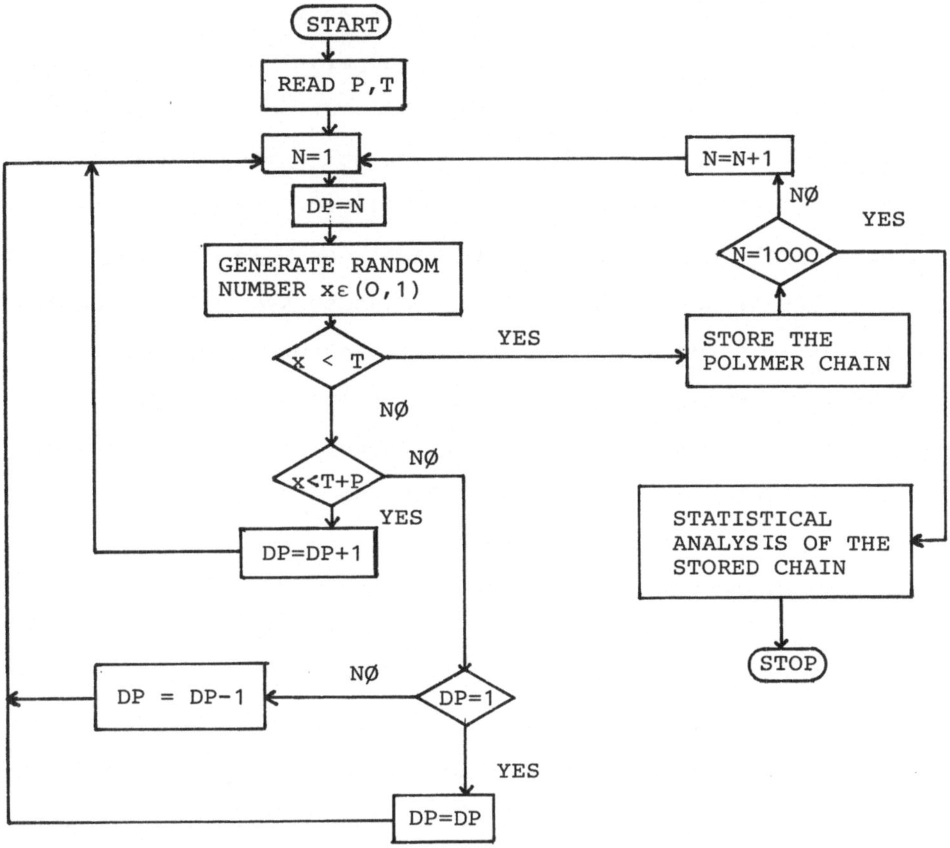

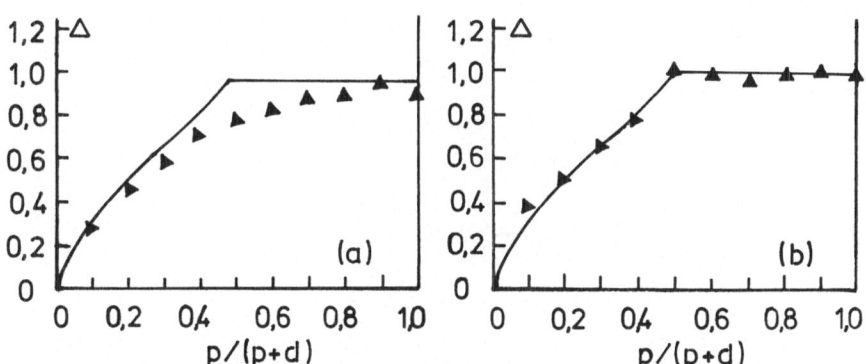

```
                    ┌───────────┐
                    │   START   │
                    └───────────┘
                          │
                    ┌───────────┐
                    │ READ  P,T │
                    └───────────┘
                          │
         ┌──────────►┌───────────┐◄──────────────────────┌───────────┐
         │           │   N=1     │                        │  N=N+1    │
         │           └───────────┘                        └───────────┘
         │                 │                                    ▲
         │           ┌───────────┐                              │  NØ
         │           │   DP=N    │                           ◇─────────◇
         │           └───────────┘                         ◇  N=1000   ◇   YES
         │      ┌────────────────────┐                      ◇─────────◇
         │      │ GENERATE RANDOM    │                           ▲
         │      │ NUMBER xε(0,1)     │                  ┌──────────────────┐
         │      └────────────────────┘                  │  STORE THE       │
         │              │          YES                   │  POLYMER CHAIN   │
         │          ◇───────◇──────────────────────────►└──────────────────┘
         │         ◇  x < T  ◇
         │          ◇───────◇
         │              │ NØ
         │              │        NØ
         │          ◇───────◇─────────────────┐      ┌──────────────────┐
         │         ◇ x<T+P  ◇                 │      │  STATISTICAL     │
         │          ◇───────◇                 │      │  ANALYSIS OF THE │◄──
         │              │ YES                 │      │  STORED CHAIN    │
         │        ┌──────────┐                │      └──────────────────┘
         │        │ DP=DP+1  │                │               │
         │        └──────────┘                │          ┌──────────┐
         │                                    │          │   STOP   │
         │                          NØ        ▼          └──────────┘
         │   ┌──────────┐◄─────────────◇───────◇
         │   │ DP = DP-1│              ◇ DP=1  ◇
         │   └──────────┘               ◇───────◇
         │                                  │ YES
         │                            ┌──────────┐
         └────────────────────────────│  DP=DP   │
                                       └──────────┘
```

FIGURE 6. Monte Carlo computation of the molecular weight
distribution: the flow chart of the program

FIGURE 7. Polydispersity index: Δ : Monte Carlo; —— :
analytic.
a) t = 0.1; b) t = 0.001.

hetero-cooperative (auto-inhibitory) reactions.

A prototype example of a cooperative reaction is the periodate oxidation of polysaccharides. The periodate ion converts each vicinal diol into aldehyde group which may form a hemiacetal with the closest hydroxyl group on adjacent unoxidized residues, thereby protecting them against oxidation. This blocking effect limits the final oxidation of alginates to 44%. Other reactions with cooperative effects are: periodate oxidation of amylose, quaternization of 4-vinyl pyridine, chlorination of polyethylene, epoxidation of natural rubber, hydrolysis of poly (methyl methacrylate) etc.

Monte Carlo method provides the simples route to get informations on sequence distribution in the intermediate copolymers of the homopolymers poly (A) and poly (B):

$$-A-A-A-A-A-A- \longrightarrow -A-B-A-B-B-A- \longrightarrow ... \longrightarrow -B-B-B-B-B-B-$$

Because during a polymer analogous (or, modification) reaction the environments of the reactive mers change, the reactivities of these mers may thereby change too. The simplest way in evaluating the reactivities is to study the changes in the sequence distributions.

The first Monte Carlo works on reactivity in polymer analogous reactions are due to Klesper et al[91-93]. These authors have assumed that the reaction rates are influenced by only the two nearest neighbours.

$$\sim A-A-A \sim \xrightarrow{k(AAA)} \sim A-B-A \sim$$

$$\sim A-A-B^+ \sim \xrightarrow{k(AAB^+)} \sim A-B-B^+ \sim$$

$$\sim B-A-B \sim \xrightarrow{k(BAB)} \sim B-B-B \sim$$

The superscript + indicates that $\sim AA-B^+\sim$ includes both $\sim A-A-B \sim$ and $\sim B-A-A \sim$. The reaction probabilities p(AAA),

$p(AAB^{+})$ and $p(BAB)$ differ from the corresponding reaction rate constants $k(XAY)$ by a common factor:

$$p(XAY) = \frac{1}{k_{(AAA)} + k_{(AAB^{+})} + k_{(BAB)}} \ k_{(XAY)} \ .$$

The algorithm used by Klesper et al. is:

1. Consider a model chain consisting of 100 or 1000 mers.

2. Select randomly the A -mer to be converted to a B-mer. This selection is performed using the first two or three digits of a uniform random number between O and 1.

3. Test if A is converted to B: if an unused uniform random number $x\varepsilon(O,1)$ lies within the range O to the appropiate reaction probability, the particular A-mer is considered to have reacted, otherwise it is left unreacted.

4. Repeat the steps 2 and 3 until the desired conversion is reached.

5. Analyse statistically the generated polymer chain.

In order to minimize the statistical fluctuation of the relative frequences of monads and sequences, several runs are performed, and the relative frequences resulted from the same number of used random numbers are averaged.

Harwood et al.[94] modified the procedure of Klesper et al. introducing "the decreasing - table technique" (one increases in this way the efficiency of the calculations). In this technique, the unreacted A sites are selected by random selection of position in a table containing the position number of unreacted A sites. The position numbers of reacted sites are removed periodically from the table, reducing in this way the amount of computation.

Harwood et al. developed the following procedure to determine the rate constants $k(XAY)$ from given triad distribution - conversion data:

1. Using the Monte Carlo model and various sets of k values, one computes triad distribution at the experimental conversions.

2. The calculated triad distributions are compared with corresponding experimental data. One selects the $k_{(XAY)}$ values for which the best agreement is obtained (the selection is performed using statistical criteria; for details see ref. 94).

As an example[94], Table 4 compares the relative reactivities evaluated by the method described above with the reactivities used to generate the triad distribution analysed.

For other Monte Carlo simulations within this class of reactions see ref. 95, 96 and 109. The analytical treatment of these reactions is discussed in ref. 110.

Table 4. Relative reactivities from triad distributions
at various conversions.

Actual reactivities			Relative reactivities			Relative reactivities found by Monte Carlo		
$k_{(AAA)}$	$k_{(AAB)}$	$k_{(BAB)}$	$k'_{(AAA)}$	$k'_{(AAB)}$	$k'_{(BAB)}$	$k'_{(AAA)}$	$k'_{(AAB)}$	$k'_{(BAB)}$
1	1	1	0.33	0.33	0.33	0.31	0.34	0.34
1	1	10	0.08	0.08	0.83	0.10	0.10	0.80
1	2	4	0.14	0.29	0.57	0.14	0.30	0.55
4	2	1	0.57	0.29	0.14	0.59	0.26	0.14
35.7	6	1	0.84	0.14	0.02	0.83	0.14	0.03
1	6	35.7	0.02	0.14	0.84	0.02	0.13	0.85

1.4.5. <u>Reactivity in binary irreversible copolymerization</u>

Motoc proposed[97] to interpret the sequence distribution in copolymers as a chemical dictionary:

$$
\begin{array}{ccccc}
\text{mers} \longrightarrow \text{blocks} & (-M_1-)_1, & (-M-)_{n>1} & \longrightarrow & \text{macromolecule} \\
\downarrow & \downarrow & \downarrow & & \downarrow \\
\text{letters} \longrightarrow \text{punctuations,} & \text{words} & \longrightarrow & \text{sentence, grammar} \\
& & & & \text{(chemical dictionary)}
\end{array}
$$

This author considers that the diagramm (19) is commutative(i.e., one can reverse the direction of the crosswise arrows). In this case, the chemical dictionary can be used to determine the reactivity ratios which are "memorized" by the macromolecule, and are "written" in the corresponding sentence.

It seems to be right to interpret the blocks $(-M_1-)_1$ and $(-M_2-)_1$ as punctuations, and the blocks $(-M_1-)_{n \geqslant 2}$ and $(-M_2-)_{n \geqslant 2}$ as words. Our Monte Carlo simulations[88,89] of the binary irreversible copolymerization pointed out that the monads are very sensitive to the perturbations of the r_1 and r_2 values, while the longer blocks are stable. Obviously, the meaning of the sentence lies in the blocks with $n \geqslant 2$, and the variability of the monads can be regarded as spelling mistakes.

In order to compute the reactivity ratios we developed[97, 101, 102] the following strategy: one computes all the r_1, r_2 pairs which generate macromolecules with the same composition F_1 (or F_2), for a given feed composition f_1 (of f_2). One selects the appropiate r_1, r_2 values using one of the following two methods:

1) one takes into account the informations concerning the sequence distribution (if available), knowing that the macromolecules discriminate well in the sequence distribution space[100].

2) one considers the compositon data in the manner detailed below. One proceeds considering N pairs of compositions $(f,F)_I$,

$I = 1, 2, \ldots, N$, and the sets S_I are computed:

$$S_I = \{(r_1, r_2)_{I,J}\} \quad , \qquad J = 1, 2, \ldots, N_I$$

I indexes the different feed compositions used to synthesize the considered copolymer (the other conditions are fixed), and J indexes the reactivity ratios pairs which gave the same copolymer composition.

Due to the error of the experimental compositon and to the simplifications of the mathematical model (see subsection 2.2), we must consider, as definition, that:

$$(r_1, r_2)_I = (r_1', r_2')_I, \text{ if and only if}$$

$$r_1 \pm \alpha_1 = r_1', \text{ and } r_2 \pm \alpha_2 = r_2', \ 0 \leqslant \alpha_1, \alpha_2 \leqslant a \qquad (29)$$

The $(r_1, r_2)_I$ pair corresponds to $(f, F)_I$, and the $(r_1', r_2')_I$, one, to $(f, F)_{I'}$. The value of a is choosen such as to assure the desired accuracy of the reactivity ratios of the considered monomers.

Using the definition (29), one performs the intersections:

$$\bigcap_{I=1}^{N} S_I = \{(r_1, r_2)_K\} \quad , \ K = 1, 2, \ldots, L \qquad (30)$$

where K indexes the reactivity ratios pairs which obey relation (29). Relation (30) is justified because the monomer reactivity depends on its chemical identity, so the desired r_1, r_2 values must exist in every set S_I, $I = 1, 2, \ldots, N$ (i.e., the set $\bigcap S_I$ is not empty). Because both experimental data and the computing method are mortaged by errors and simplifications, respectively, $L > 1$ (i.e., the set $\bigcap S_I$ has $L > 1$ elements, not only one).

The desired r_1, r_2 values are obtained as average of the r_1's and, respectively, r_2's common to the N sets S_I:

$$r_1 = \frac{1}{L} \sum_{K=1}^{L} r_1, \quad \text{and} \quad r_2 = \frac{1}{L} \sum_{K=1}^{L} r_2.$$

If the standard deviation of the resulted r_1, r_2 values is not satisfactory, one should use in the relations (20) a smaller value a' < a.

The method described above has been implemented through the algorithm:

1. Input data: N (polymerization degree), NA and NB (the number of M_1- and M_2 - mers in the macromolecule with polymerization degree N), C(the ratio f_2/f_1).

2. Consider a pair of α and β values. α and β are computed according to the relations:

$$\alpha = r_1/(r_1+C), \quad \text{and} \quad \beta = r_2/(r_2+C^{-1}).$$

3. Simulate the macromolecule corresponding to α and β values. Characterize the resulted macromolecule by composition (the number of M_1- and M_2- mers are denoted by AI and BI, respectively) and sequence distribution.

4. If AI $\pm$ 5 = NA and BI $\pm$ 5 = NB the informations concerning the macromolecule are stored, otherwise they are rejected.

5. β is advanced with the amount 10/N, α is kept fixed, and the algorithm is continued with the point (3). When β = 1, the algorithm is continued with the point (6).

6. α is advanced with the amount 10/N, β is set 1/N, and the algorithm is continued with the point (3). When α = 1, the computing procedure is stopped.

7. Output data: the list of computed r_1, r_2 values and corresponding sequence distribution data.

The flow chart of the program is shown in Figure 8. The program is termed MEMORY-7, its listing being included within the MEMORY program given in the section 4.1.

The program efficiency is 20 minutes/one input data set.

The method described here to compute the reactivity ratios satisfies the condition pointed out by Bechnken[104] and Tidwell. and Mortimer[105]; i.e., the method does not transform the error structure in the observed copolymers composition.

MEMORY-7 has been used to study the copolymerization of several concrete comonomer pairs. The obtained results are summarized below.

A) Copolymerization of methyl methacrylate (M_1) and chloroprene (M_2).

The experimental data are taken from ref. 103 (with 85%M_1 in the feed there resulted a copolymer with 40%M_1, mole - %).

MEMORY-7 furnished[97] two sets of reactivity ratios centered around $r_1 = 0.005$ and $r_2 = 2.774$. (40.4%M_1 in macromolecule), and $r_1 = 0.058$ and $r_2 = 5.601$ (39.5%M_1 in macromolecule). The computed sequence distributions are displayed in Table 5.

TABLE 5. Methyl methacrykate/chloroprene copolymer:
sequence distribution

k	r_1=0.005, r_2=2.774		r_1=0.058, r_2=5.601		r_1=0.08, r_2=5.1	
	$n_1(k)$	$n_2(k)$	$n_1(k)$	$n_2(k)$	$n_1(k)$	$n_2(k)$
1	379	254	224	145	211	146
2	10	84	51	72	52	75
3	–	36	18	43	23	40
4	–	11	2	15	5	16
5	1	3	–	6	–	6
6		–	–	4	–	5
7		1	2	3	2	1
8				1		–
9				1		1
10				–		–
...						
16				1		1

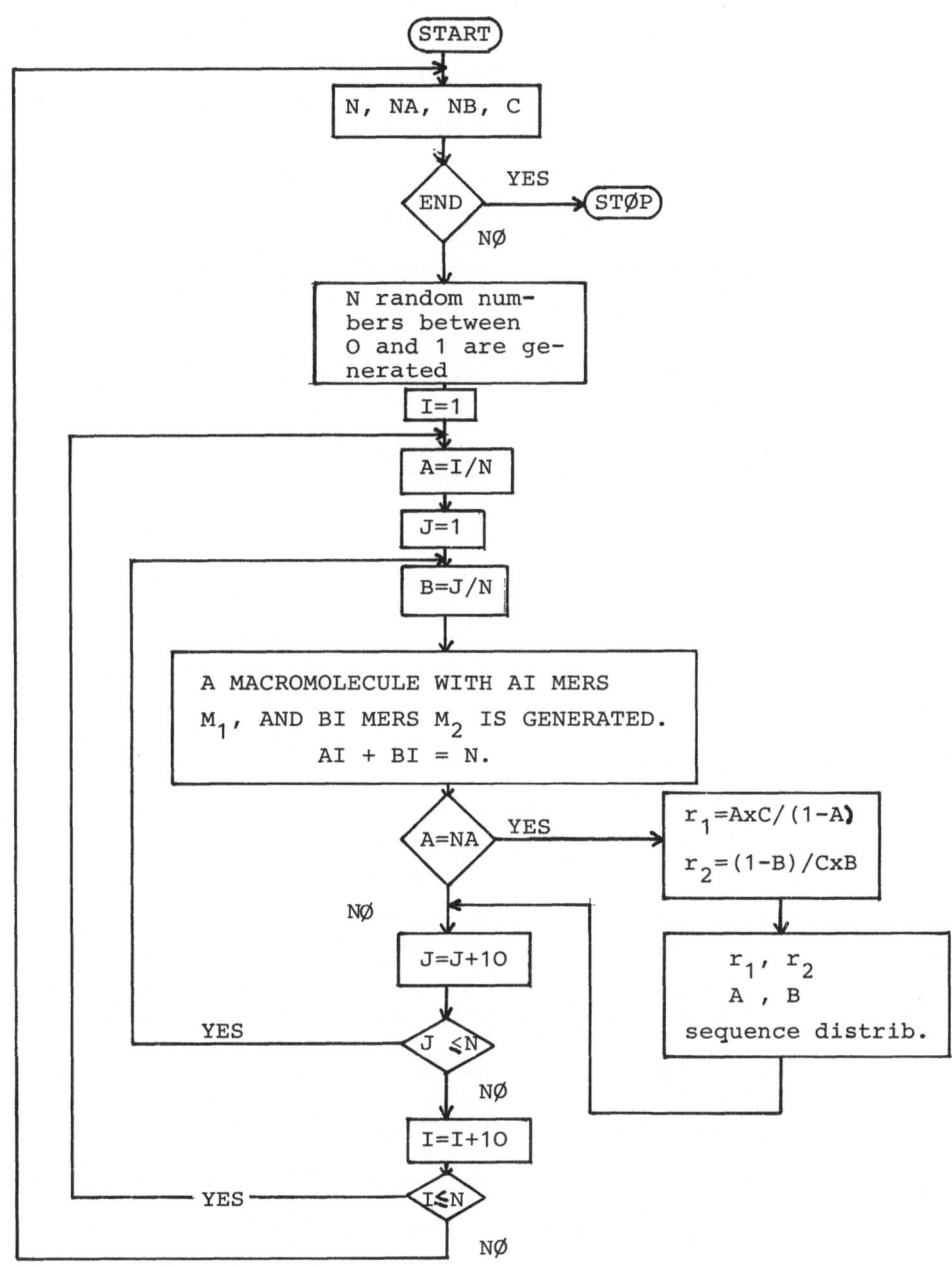

FIGURE **8.** The flow chart of MEMORY-7 program

n(k) stands for the number of sequences of length k; subscripts 1 and 2 indicate the nature of the sequence.

Using the sequence distribution data reported by Ebdon[103], one selects the values r_1 = 0.058 and r_2 = 5.601. These values compare well with the values obtained by Braun et al., namely r_1 = 0.08 and r_2 = 5.1 (the Monte Carlo sequence distribution corresponding to Braun's reactivity ratios is also shown in Table 5).

B) Copolymerization of acrylic acid (M_1) and methyl acrylate (M_2).

The experimental data displayed in Table 6 are taken from ref. 107.

Table 6.   Acrylic acid/methyl acrylate copolymers: MEMORY-7 input data (N=1000).

No.	C	NA	NB
1.	19.0/81.0	857	143
2.	27.6/72.4	772	228
3.	28.4/71.6	756	244
4.	41.1/58.9	615	385
5.	59.0/51.0	561	439
6.	54.2/45.8	491	509
7.	66.0/34.0	411	589
8.	69.5/30.5	326	674

Table 7 collects the resulted[102] S_I sets, I = 1,2,...,8. Because informations regarding the sequence distribution in these copolymers are not available till now, the set $\cap S_I$ has been computed. The elements of the $\cap S_I$ are in the area marked by dashed lines in Table 7. Averaging these values, one gets:

$$r_1 = 17.857/13 = 1.374, \quad \text{and} \quad r_2 = 13.492/13 = 1.038 .$$

These values compare well with those of Eldridge and Treloar[107] (determined by Fineman – Ross method, and checked by the methods of Tidwell-Mortimer and Kelen-Tüdös), namely: r_1 = 1.4 and r_2 = 1.0.

In order to argue the statement that macromolecules discriminate well in the sequence distribution space, the corresponding Monte-Carlo data for two copolymer compositions are collected[102] in Table 8.

Table 7. Acrylic acid/methyl acrylate copolymers: the resulted sets of copolymerization ratios pairs (copolymer no. corresponds to Table 6)

Copolymers No.	1		2		3	
	r_1	r_2	r_1	r_2	r_1	r_2
	1.076	0.173	0.894	0.051	0.809	0.075
	1.153	0.806	0.938	0.106	0.847	0.246
	1.241	1.175	0.985	0.256	0.887	0.341
	1.340	1.490	1.036	0.354	0.930	0.407
	1.453	1.733	1.091	0.459	0.976	0.441
	1.584	2.493	1.150	0.652	1.025	0.513
	1.737	2.830	1.214	0.779	1.078	0.588
	1.917	3.617	1.283	0.965	1.135	0.707
	2.135	4.419	1.359	1.066	1.196	0.836
	2.401	5.404	1.443	1.286	1.263	0.976
	2.735	5.404	1.534	1.742	1.335	1.127
	3.165	7.228	1.636	1.892	1.415	1.293
	3.741	10.387	1.748	2.053	1.501	1.674
	etc.		1.874	2.317	1.597	1.973
			2.016	2.226	1.702	2.227
			2.177	3.325	1.819	2.318
			2.361	3.198	1.950	2.614
			2.574	4.262	2.098	3.069
			2.822	4.850	2.265	3.766
			3.116	5.549	2.457	4.096
			3.469	6.391	2.678	4.662
			etc.		etc.	

Table 7 (continued)

4		5		6		7		8	
r_1	r_2	r_1	r_2	r_1	r_2	r_1	r_2	r_1	r_2
0.361	0.043	0.212	0.009	0.090	0.114	0.126	0.276	0.174	0.556
0.507	0.251	0.414	0.212	0.244	0.223	0.242	0.328	0.314	0.629
0.701	0.426	0.520	0.275	0.397	0.343	0.373	0.403	0.374	0.656
0.822	0.555	0.617	0.345	0.440	0.414	0.488	0.420	0.405	0.648
0.926	0.703	0.758	0.444	0.510	0.473	0.519	0.420	0.470	0.713
1.008	0.768	0.822	0.488	0.728	0.635	0.619	0.490	0.538	0.744
1.143	0.951	0.965	0.558	0.860	0.717	0.797	0.579	0.646	0.848
1.302	1.034	1.132	0.782	0.972	0.746	0.836	0.602	0.981	0.972
1.423	1.266	1.388	0.996	1.142	0.912	0.960	0.627	1.077	1.019
1.490	1.318	1.447	0.966	1.287	1.071	1.004	0.653	1.127	1.019
1.717	1.486	1.643	1.076	1.395	1.116	1.145	0.738	1.463	1.123
1.996	1.744	1.960	1.319	1.452	1.211	1.411	0.837	1.590	1.180
2.222	2.141	2.149	1.555	1.512	1.211	1.560	0.953	1.872	1.243
2.488	2.650	2.364	1.691	1.641	1.316	1.799	0.996	2.029	1.382
2.641	2.770	2.611	1.765	1.710	1.316	1.949	1.041	2.288	1.461
2.994	3.492	3.059	2.202	2.024	1.562	2.111	1.090	3.033	1.744
3.985	4.771	3.426	2.417	2.113	1.562	2.382	1.196	3.579	2.128
4.711	6.070	4.123	3.278	2.526	1.872	2.584	1.255	4.443	2.467
etc.		etc.		2.774	1.962	2.924	1.318	5.092	2.674
				2.911	2.059	3.785	1.537	5.889	3.188
				3.386	2.273	3.959	1.622	6.519	3.515
				etc.		4.341	1.715	etc.	
						etc.			

Table 8. Macromolecules discrimination in sequence distribution space[*]

k	Copolymer No. 1				Copolymer No. 2					
	a		b		a		b		c	
	$n_1(k)$	$n_2(k)$	$n_1(k)$	$n_2(k)$	$n_1(k)$	$n_2(k)$	$n_1(k)$	$n_2(k)$	$n_1(k)$	$n_2(k)$
1.	24	135	21	110	264	174	126	81	14	7
2.	18	6	13	20	30	74	52	48	10	6
3.	19		14	1	3	32	21	29	3	1
4.	12		12		–	12	7	18	5	5
5.	10		12		–	4	1	6	3	4
6.	14		15		1	1	1	9	5	2
7.	8		5			–	1	6	4	2
8.	2		6			1		4	4	2
9.	3		2					2	2	1
10.	8		6					3	–	4
11.	4		4					–	2	1
12.	2		3					1	3	5
13.	5		5					1	–	–
14.	2		1						1	1
15.	5		5						2	4
16.	1		3						1	2
17.	2		2						1	1
18.	1								–	2
19.									1	1

[*] 1-a: $r_1 = 1.07$; $r_2 = 0.173$ 2-a: $r_1 = 0.24$; $r_2 = 0.33$
    1-b: $r_1 = 1.24$; $r_2 = 1.170$ 2-b: $r_1 = 1.41$; $r_2 = 0.84$
                                      2-c: $r_1 = 10.27$; $r_2 = 5.15$

The sequence distributions were computed by means of MEMORY-3 program, using the reactivity ratios furnished by MEMORY-7 type calculations.

      C). Copolymerization of methyl methacrylate (M_1) and chloroprene (M_2). Resumption.

We shall resume[36] the calculation of reactivity ratios for copolymerization of methyl methacrylate and chloroprene using the composition data[103] collected in Table 9.

Table 9. Methyl methacrylate/chloroprene copolymers:
          composition data.

No.	C	NA	NB
1.	2/98	860	140
2.	3/97	770	230
3.	4/96	710	290
4.	5/95	640	360
5.	10/90	480	520

Several elements of the S_I sets, $I = 1,2,...,5$, are displayed in Table 10. The elements of the $\cap S_I$ set are in the area
marked by dashed lines in Table 10. Averaging these r_1, r_2 values,
one obtains:

$$r_1 = 0.587/8 = 0.073, \text{ and } r_2 = 46.819/8 = 5.852$$

$r_1 = 0.073$, $r_2 = 5.852$ are in good agreement with both previously
computed reactivity ratios ($r_1 = 0.058$, $r_2 = 5.601$) and the values
determined by Braun et al.[106] ($r_1 = 0.08$, $r_2 = 5.1$).

For a short review of MEMORY-7 type computations one may
consult ref. 108.

1.4.6. Step-growth polymerization

The first Monte Carlo works dealing with polymerizations
which follow a step-growth mechanism are due to Glasser and Glasser
[111], Falk and Thomas[112], Siling et al.[113], and Motoc et al.[123,124].
Glasser and Glasser[111] simulated the naturally occuring polymerization of lignin. Their model is relatively incomplete due to
the assigment of the reaction probabilities.

Falk and Thomas[112] dealt with the self-condensation of f-
functional monomer RA_f, where the functional groups A may combine

Table 10. Methyl methacrylate/chloroprene copolymers: the sets S_I and $\cap s$ (copolymer no. corresponds to Table 10).

Copolymer no.	1		2		3		4		5	
r_1, r_2 values	r_1	r_2	r_1	r_2	r_1	r_2	r_1	r_2	r_1	r_2
	0.094	1.463	0.068	0.633	0.057	0.732	0.036	0.585	0.000	0.989
	0.100	5.994	0.072	0.976	0.061	1.816	0.042	1.915	0.004	1.101
	0.108	12.174	0.075	1.326	0.064	2.394	0.045	3.164	0.005	1.333
	0.117	16.246	0.079	3.996	0.067	3.309	0.055	4.568	0.006	1.453
			0.083	5.274	0.070	4.290	0.058	4.869	0.023	2.984
			0.088	6.176	0.076	4.632	0.060	5.179	0.043	5.041
			0.092	8.116	0.080	5.710	0.063	5.496	0.053	6.228
			0.097	9.703	0.083	7.277	0.071	7.584	0.060	6.762
			0.103	12.021	0.091	8.563	0.084	8.765	0.063	7.043
			0.116	16.753	0.096	9.481	0.095	11.962	0.068	7.334
					0.100	9.959	0.104	14.135	0.088	8.613
					0.105	11.477	0.126	17.311	0.103	10.523
					0.117	14.327			0.112	11.408
									0.121	12.378

to result highly branched and/or cyclized polymers. This work has shown in an elegant manner that Monte Carlo simulation can resolve the differences between two analytical treatments (in the present case, the treatments of molecular weight distribution due to Flory[114] and Masson et al.[115, 116,] respectively).

Siling et al.[113] computed the isomeric composition (Staudinger[117]) of macromolecules resulted by phenol/formaldehyde polycondensation. The topic is important because the isomeric composition conditions the resin properties[118]. These authors dealt with the reaction scheme(I). k stands for the reaction rate, and subscripts o and p for ortho and para.

At a given reaction step there results the ortho-methylol derivative with the probability W_o, and the para-methylol derivative with the probability W_p. The relative probability W_o/W_p is given by the relation:

$$\frac{W_o}{W_p} = \frac{k_o H_o}{k_p H_p} .$$

The probabilities to result an o-o, o-p or p-p methylene bridge are W_{oo}, W_{op} and W_{pp}:

$$\frac{W_{oo}}{W_{op}} = \frac{k_{oo} H_o}{k_{op} H_p} , \text{ and } \frac{W_{pp}}{W_{op}} = \frac{k_{pp} H_p}{k_{op} H_o} ,$$

where H_o and H_p denote the number of available positions for ortho and para attack in the considered reaction step.

One has started the simulations with a system consisting of N phenol molecules and 0.8 N formaldehyde molecules.

If the reaction occurs at 100^oC and pH = 1, k_o/k_p = 0.450, k_{oo}/k_{op} = 0.115 and k_{po}/k_{pp} = 0.105 (according to refs. 119-121).

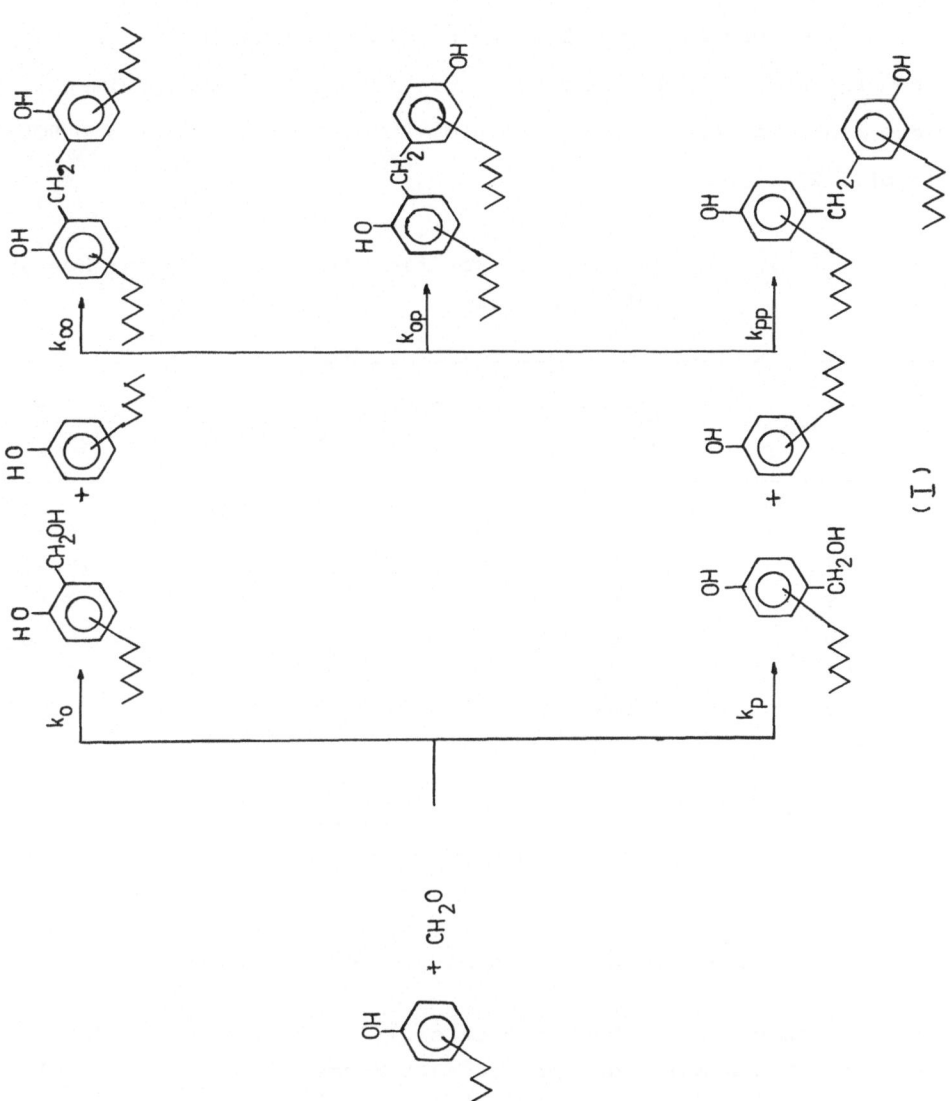

(I)

The obtained results are displayed in Table 11. The computed isomeric composition is in good agreement with the experimentally determined values, namely[122]: (o,o) : 10 + 20%; (o,p) : 5 + 60%; (p,p) : 25 + 30%.

Table 11. Isomeric composition of Novolac resin: Monte Carlo results

$N/10^3$	mole-% of methylene bridges		
	(o,o)	(o,p)	(p,p)
95	19.76	52.11	28.13
97	19.56	52.19	28.23
99	19.64	51.91	28.45
100	19.70	52.13	28.16
102	19.65	52.11	28.24
105	19.69	52.09	28.21
200	19.72	52.08	28.20
Mean values	19.68	52.09	28.23

Polyaddition reaction:

$$\text{diisocynate} + \text{diol}_1 + \text{diol}_2 \longrightarrow \text{polyurethane}$$

has been investigated by Motoc et al.[123, 124] using test data. The performed simulations proved that Monte Carlo is an useful method to evidence the features of the polyurethane at the microscopic level.

1.4.8. Inhomogeneity in copolymers

The statistical stationarity with regard to the growing active species is not realized when the polymerization degree is not sufficiently large. In consequence, the effect of the first mer of the macromolecule can not be neglected, and, accordingly, the sequence of two or more mers is not identical in all the chains. Thus, inhomogeneity with respect to chemical composition occurs.

Tsuchida et al. [133-136] investigated this topic from an experimental point of view in the case of synthetic cooligomers. Inhomogeneity is a very important aspect of polysaccharide chemistry[137],

The first attempt to treat theoretically the composition inhomogeneity is due to Fueno and Furukawa[138]. For the copolymer composition (assuming an ultimate effect mechanism) they derived the equation:

$$P(M_1)^n = F_{M_1} + (X_{M_1} - F_{M_1}) \lambda^{n-1}, \qquad (31)$$

where n stands for the polymerization degree, and

$$F_{M_1} = P_{M_2 M_1} / (P_{M_1 M_2} + P_{M_2 M_1}),$$

$$\lambda = 1 - (P_{M_1 M_2} + P_{M_2 M_1}).$$

$P_{M_1 M_2}$ and $P_{M_2 M_1}$ are the conditional probabilities that the growing species ending in M_1 or M_2 select the M_2 or M_1 monomers in the propagation step. X_{M_1} is the probability that M_1 is the first mer of the macromolecule.

If the polymerization degree is large, the term $(X_{M_1} - F_{M_1}) \cdot \lambda^{n-1}$ is nearly equal to zero, and F_{M_1} (Mayo-Lewis equation) defines the composition of the copolymer. For small values of n, the equation (22) evidences the compositon inhomogeneities.

The first Monte Carlo investigation of chemical inhomogeneity in copolymers is due to Smidsrod and Whittington[132]. The parameters needed for the simulation are the probabilities $p(M_1M_1)$, $p(M_1M_2)$, $p(M_2M_1)$ and $p(M_2M_2)$, if the first two mers of the macromolecule are M_1M_1 etc., and the conditional probabilities $p(M_I/M_JM_K)$ $I,J,K = 1,2$, which express the probability that the monomer M_I is selected given that the ultimate mer is M_K and the penultimate mer is M_J. The computer program developed by Smidsrod and Whittington is a simplifyed version of our MEMORY-5 program and detailed description of the latter program is given in the subsection 2.3. Typical results are shown in Figure 9.

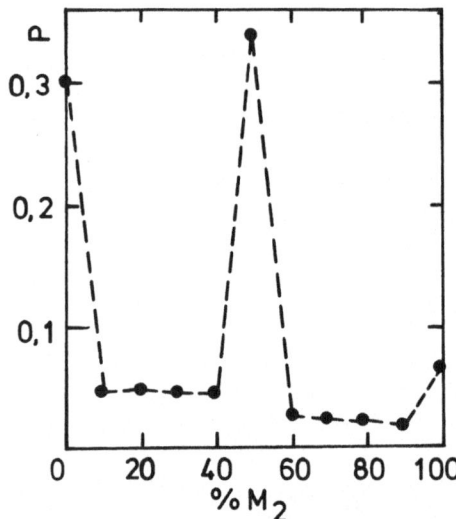

Figure 9. Composition distribution for polymers of polymerization degree n=10. $p(M_2M_1)$ = $p(M_1M_2)$ = 0.2369, $p(M_2M_2)$ = 0.1104, $p(M_1M_1)$ = 0.4157, $p(M_1/M_1M_1)$ = 0.9677, $p(M_2/M_2M_2)$ = 0.9375, $p(M_1/M_2M_1)$ = 0.0566, $p(M_2/M_1M_2)$ = 0.0291.

One may easily observe that for long polymeric chains, the composition distribution is a δ – function at 34.7% M_2.

1.4.9. Conformation and sequence distribution

The advantages of the Monte Carlo approach over analytical methods for the computation of the sequence distribution are i) the easiness to account for various polymerization mechanisms; ii) its ability to provide approximate data for small as well as high conversions; iii) the possibility to visualize the features of the polymer chain.

The Monte Carlo algorithms developed to treat the sequence distribution and polymer conformation are reviewed in the chapters 2 and 3, respectively.

The configuration of polymer molecules, i.e. their spatial geometries, can be treated by analytical methods only if the molecules are unperturbed. That means: no forces other than short-range interferences (constant bond lengths and valence angles, hindered rotations about the bonds) between the structural units are taken into account. Therefore the results of such a treatment are valid only for that special case. But there do exist general van der Waals interactions between structural units even if they are remote from one another along the chain. It is these interactions which give rise to the so-called excluded-volume effect. Its influence on the configuration of macromolecules has been the subject of numerous theoretical investigations during the last three decades. Yet this problem could not be solved satisfactorily, due to mathematical difficulties which are enormous compared with the unperturbed case.

These difficulties require simplifications whose effect on the final result can hardly be estimated. In many cases it is not possible to decide from experiment whether a theory gives correct results or not, because the equations given by theory generally contain parameters which are not directly observable quantities.

Therefore, when comparing theory with experiment these parameters
can only be treated as adjustable items, although they have
a well-defined meaning. In this connection the Monte Carlo method
is a powerful tool not only in testing theories, but also in giving
deeper insights into the configurational behaviour of macromolecules
under various conditions.

Appendix

The elements of a stochastic matrix $P = (p_{ij})$ must satisfy the
conditions

$$p_{ij} \geqslant 0 \qquad\qquad \sum_j p_{ij} = 1$$

According to the definition given in Section 1.3.2 we have

$$p_{ij}^{(n)} = (P^n)_{ij}$$

P^n can be calculated by diagonalizing P.

$$P\ R = R\ \Lambda \tag{1}$$

with $\Lambda = (\lambda_i\ \delta_{ij})$, a diagonal matrix with the eigenvalues
λ_i of P as elements, and R, a matrix composed of the eigenvectors
belonging to the respective λ_i. From (1) follows

$$P = R\Lambda R^{-1} = R\Lambda L \quad (L \equiv R^{-1})$$

and

$$p_{ij}^{(n)} = (R\Lambda^n L)_{ij} = \sum_{st} r_{is}\ \lambda_s^n\ \delta_{st}\ l_{tj} = \sum_t r_{it}\ \lambda_t^n\ l_{tj}$$

It is known from the theory of stochastic matrices [140] that the
largest eigenvalue is non-degenerate and equals unity, i.e.
$\lambda_1 = 1$, the corresponding eigenvector being $\vec{r}_1^T = (1,1,\ldots,1)$. For
large n the term with λ_1 therefore dwarfs all others. In this limit
$p_{ij}^{(n)}$ is given by

$$\lim_{n\to\infty} p_{ij}^{(n)} = r_{i1}\ \lambda_1^n\ l_{1j} = l_{1j} = \pi_j$$

l_{1j} is a component of the row vector $\vec{l}_1^T$, and can be evaluated by
considering the eigenvalue problem

$$\vec{l}_1^T\ P = \lambda_1\ \vec{l}_1^T \quad \text{or in another notation}$$

$$\sum_i l_{1i} p_{ij} = l_{1j} \quad \text{which is equivalent to} \quad \sum_i \pi_i p_{ij} = \pi_j\ .$$

References

1. C.W. Churchman, in "Symposium on Monte Carlo Methods", A.C. Hoggart and F.E. Balderton eds., S.W.Publ.Comp., Cincinnati, 1963.

2. S.Wold and M.Sjöström, in "Chemometrics: Theory and Applications", B.R. Kowalski ed., ACS Symposium Series, No. 52, Washington, 1977.

3. a. N.B.Chapman and J. Shorter eds., "Advances in Linear Free Energy Relationships", Plenum, New York, 1972; b) W. Purcell, G. E. Bass and J.M. Clyton, "Strategy of Drug Design. A Guide to Biological Activity", Wiley, New York, 1973.

4. M.E. Drummond Jr., "Evaluation and Measurement Techniques for Digital Computer Systems", Prentice-Hall, Englewood Cliffs, N.J., 1973.

5. D.J. Newman, Aust. J.Phys., $\underline{31}$, 489 (1978).

6. P.A.D. de Maine, Comput. Chem., $\underline{2}$, 53 (1978).

7. R.A. Merikallio and F.C. Holland, in "Simulation Design of a Multiprocessing System", Proc. FJCC, vol. 33, part 2, 1968, p. 1399 - 1410.

8. O.J. Campbell and W.J. Heffner, ibid, vol. 33, part 1, 1968, p. 903 - 914.

9. M. Greenberger, "A New Methodology for Computer Simulation", MIT, Cambridge , AD 609288, 1964.

10. T.H. Naylor et al., "Computer Simulation Techniques", Wiley, New York, 1966.

11. G.W. Evans II, G.F. Wallance and G.L. Sutherland, "Simulation Using Digital Computers", Prentice-Hall, Englewood Cliffs, N.J., 1967.

12. F.F. Martin, "Computer Modelling and Simulation", Wiley, New York, 1968.

13. C. Fevrier, "Le Simulation de Systemes", Dunod, Paris, 1972.

14. J.W. Mayne, J.Can. Operational Res.Soc., $\underline{4}$, 144 (1966).

15. R. von Mises, Math. Z., $\underline{5}$, 52 (1919).

16. O. Onicescu, "Nombres et Systems Aleatoires", Editions Eyrolles, Paris, 1964.

17. M.D. MacLaren and G. Marsaglia, J. Assoc. Comput. Mach., <u>12</u>, 83 (1965).

18. B. Jansson, "Random Numbers Generators", Almqvist and Wiksell, Stockholm, 1966.

19. E.R. Sowey, Internat. Statist. Rev., <u>40</u>, 355 (1972).

20. O. Dragomir-Filimonescu, Proc. Fifth National Symp. Information Sci., Cluj-Napeca, 1979 (in Roumania).

21. N. Metropolis and S.M. Ulan, J.Am. Statist. Assoc., <u>44</u>, 335 (1949).

22. H. Niederreiter, Bull. Am. Math. Soc., <u>84</u>, 957 (1978).

23. H. Kahn, "Applications of Monte Carlo", Rand Memo. RM-1237, 1954.

24. A. Meyer, ed., "Symposium on Monte Carlo Methods", Wiley, New York, 1956.

25. E.D. Cashwell and C.J. Everett, "Practical Manual on the Monte Carlo Method for Random Walk Problems", Pergamon, New York, 1959.

26. J.M. Hammersley and D.C. Handscomb, "Monte Carlo Methods", Wiley, New York, 1964.

27. Yu.A. Shreider, ed., "The Monte Carlo Method", Pergamon, New York, 1965.

28. K.F. Iuorno, "Simulation Using the Monte Carlo Method", Griffiss Air Force Base, N.Y., AD-410290, 1973.

29. N.P. Buslenko et al., "The Monte Carlo Statistical Trials", Fizmatgis, Moscow, 1962 (in Russian).

30. S.M. Yermakov, "The Monte Carlo Method and Relevant Problems", Nauka, Moscow, 1971 (in Russian).

31. I.M. Sobol, "Numerical Monte Carlo Methods", Nauka, Moscow, 1973 (in Russian).

32. K. Binder, ed., "Monte Carlo Methods in Statistical Physics", Springer, Berlin, 1979.

33. K.F. O'Driscoll, in "Computers in Polymer Sciences", J.S. Mattson, H.B. Mark, Jr. and H.C. MacDonald, Jr., eds., Dekker, New York, 1977.

34. G.G. Lowry, ed., "Markov Chains and Monte Carlo Calculations in Polymer Science", Dekker, New York, 1970.

35. B. Carazza, Nuovo Cimento, 7, 419 (1977).

36. I. Motoc, Math. Chem., 5, 283 (1979).

37. H. Conroy, J. Chem. Phys., 47, 5307 (1967).

38. L.D. Fosdick, SIAM Rev., 10, 315 (1968).

39. I. Motoc, Eur. Polym. J. (submitted).

40. A. Gilath, S.H. Ronel, M. Shmueli and D.H. Kohn, J. Appl. Polym. Sci., 14, 1491 (1970).

41. F.P. Price, J. Polym. Sci., C 25, 3 (1968).

42. J.G. Topliss and R.J. Costello, J.Med. Chem.,15, 1066 (1972).

43. J.G. Topliss and R.P. Edwards, J.Med. Chem., 22, 1238 (1979).

44. A.T. Balaban, A. Chiriac, I.Motoc and Z.Simon, "Steric Fit in QSAR", Lecture Notes in Chemistry, vol. 15, Springer, Berlin, 1980.

45. G. Dashevesky and G.N. Srakisov, Mol. Phys., 27, 1271 (1974).

46. J.C. Owicki and H.A. Scheraga, J.Am. Chem. Soc., 99, 7403 (1977).

47. O. Matsucka, E. Clementi and M. Yoshimine, J. Chem. Phys., 64, 1351 (1976).

48. S. Swaminathan and D.L. Beveridge, J.Am.Chem. Soc., 101, 5832 (1979).

49. S. Okazaki, K. Nakanishi, H. Touhara and Y. Adachi, J. Chem. Phys., 71, 2421 (1979).

50. C. Pangali, M. Rao and B.J. Berne, J. Chem. Phys., 71, 2975, 2982, (1979).

51. T.L. Hill and Y. Chen, J. Chem. Phys., 69, 1126 (1978).

52. T.L. Hill and L. Stein, J. Chem. Phys., 69, 1139 (1978).

53. T.L. Hill and Y. Chen, Proc. Natl. Acad. Sci. USA, 75, 5260 (1978).

54. M.E. Craig and D.M. Crothers, Biopolymers, 6, 385 (1968).

55. I.R. Epstein, Biopolymers, 18, 2037 (1979).

56. I. Motoc, Math. Chem., 4, 113, (1978).

57. I. Motoc, R. Vancea and I. Muscutariu, Math. Chem., 5, 263 (1979)

58. I. Motoc, I. Gotaescu and R. Vancea, Nouv.J.Chim., (in press).

59. I. Motoc, S. Holban, R. Vancea and Z. Simon, Studia Biophys. (Berlin), 66, 75 (1977).

60. I. Motoc and R. Valceanu, Rev. Chim. (Bucharest), (in press).

61. J.A. Barker and R.O. Watts, Chem. Phys. Letts., 3, 144 (1969).

62. J.A. Barker and R.O. Watts, Molec. Phys., 26, 792 (1973).

63. G.N. Sarkisov and R. Dashevskii, Z. Strukturnoi Khim., 13, 199 (1972).

64. H. Popkie, H. Kistenmacher and E. Clementi, J. Chem. Phys., 59, 1325 (1973).

65. J.K. Lee, J.A. Barker and F.F. Abraham, J.Chem. Phys., 58, 3166 (1971).

66. R.O. Watts, E. Clementi and R.O. Watts, J. Chem. Phys.,61, 2550 (1974).

67. J. Fromm, E. Clementi and R.O. Watts, J. Chem. Phys., 62, 1388 (1975).

68. a)J.C. Owicki and H.A. Scheraga, J.Am. Chem. Soc., 99, 7403 (1977).
    b) S. Swaminathan and D.L. Beveridge, J.Am. Chem. Soc., 99, 8392 (1977).

69. M. Mezei, S. Swaminathan and D.L. Beveridge, J.Am. Chem. Soc., 100, 3255 (1978).

70. F.H. Stillinger and A. Rahman, J.Chem. Phys., 68, 666 (1978).

71. I.R. McDonald and M.L. Klein, J.Chem. Phys., 68, 4875 (1978).

72. M.L. Klein, I.R. McDonald and S.F. O'Shea, J. Chem. Phys., 69, 63 (1978).

73. W.L. Jörgensen, "Monte Carlo Simulation of Liquid Hydrogen Fluoride", preprint (1978).

74. G. Jacucci, I.R. MacDonald and R. Taylor, J. Phys., F8, 1121 (1978).

75. R.D. Mountain, J. Phys., F8, 1637 (1978).

76. T.Lee, J.Bishop, W.van der Lugt and W.F. Van Gunsteren, Physica, B 93, 59 (1978).

77. E. Michler, H.Hahn, and P. Schofield, J. Phys., F6, 319 (1976).

78. G. Jacucci, M.L. Klein and R. Taylor, Solid State Comm., $\underline{19}$, 657 (1976).

79. S. Miller and H.R. Clarke, J. Chem. Soc. Faraday Trans. 2, $\underline{72}$, 1372 (1976).

80. M. Dixon and M.J.L. Sangster, J. Phys., $\underline{C10}$, 3015 (1977).

81. D.J. Adams, J.Chem.Soc.Faraday Trans. 2, $\underline{72}$, 1372 (1976).

82. A.T. Hagler and J. Moult, Nature, $\underline{272}$, 222 (1978).

83. B. Robson and M.N. Jones, Nature, $\underline{272}$, 206 (1978).

84. N. Metropolis et al., J. Chem. Phys., $\underline{21}$, 1087 (1953).

85. H.L. Frisch, M. Bishop and J. Roth, J.Chem. Phys., $\underline{67}$, 1082 (1977).

86. M. Bishop, J. Roth and H.L. Frisch, J. Phys. Chem., $\underline{81}$, 2500 (1977).

87. H.L. Frisch, B. Bishop and J. Roth, in "Statistical Mechanics and Statistical Methods in Theory and Application", V. Landman ed., Plenum, New York, 1977.

88. G.G. Lowry, in "Markov Chains and Monte Carlo Calculations in Polymer Science", Dekker, New York, p. 309 - 316.

89. G.G. Lowry, Polym. Letts., $\underline{1}$, 489 (1963).

90. D.A.McQuarrie, C.J. Jachimowski and M.E. Russel, J. Chem. Phys. $\underline{40}$, 2917 (1964).

91. E. Klesper, W. Gronski and V. Barth, Makromol. Chem., $\underline{150}$, 223 (1971).

92. E. Klesper, A. Johnson and W. Gronski, Makromol. Chem. $\underline{160}$, 167 (1972).

93. E. Klesper and A.O. Johnson, in "Computers in Polymer Sciences", eds. J.S. Mattson, H.B. Mark Jr. and H.C. MacDonald Jr., Dekker, New York, 1977, ch.1.

94. H.J. Harwood, K.G. Kempf and L.M. Landoll, $\underline{16}$, 91, 109, (1978).

95. A.D. Litmanovich, N.A. Plate, O.V. Noah and V.I. Golyakov, Eur. Polym. J., Suppl., $\underline{1969}$, 517.

96. N.A. Plate and O.V. Noah, in "Advances in Polymer Science", vol. 31, ed. H.J. Cantow, Springer, Berlin, 1979.

97. I.Motoc, Math.Chem.; $\underline{8}$ (1980) - in press.

98. I. Motoc, S. Holban and D. Ciubotariu, J. Polym. Sci. Chem., $\underline{15}$, 1465 (1977).

99. I. Motoc, S. Holban and R. Vancea, J. Polym. Sci. Chem., $\underline{16}$, 1601 (1978).

100. I. Motoc, R. Vancea and S. Holban, J. Polym. Sci. Chem., $\underline{16}$, 1587 (1978).

101. I. Motoc and I. Muscutariu, J. Polym. Sci., Polym. Lett. Ed., in press.

102. I. Motoc, J. Macromol. Sci.Chem. (1980) - in press.

103. J.R. Ebdon, Polymer, $\underline{15}$, 782 (1974).

104. P.W. Bechnken, J. Polym. Sci., $\underline{A\ 2}$, 645 (1964).

105. P.W. Tidwell and G.A. Mortime, J.Polym. Sci., $\underline{A3}$, 369 (1965).

106. D. Braun, W. Brendlein and G. Mott, Eur. Polym. J., $\underline{9}$,1007 (1973).

107. R.J. Eldridge and F.E. Treloar, J. Polym. Sci. Chem., $\underline{14}$,2831 (1976).

108. I. Motoc, I. Muscutariu and S. Holban, "Preprints MakroMainz", vol. 1, eds. I. Lüdenwald and R. Weis, 1979 (26th Internat. Symp. Macromolecules).

109. O. Smidsrod, B. Larsen and T. Painter, Acta Chem. Scand., $\underline{24}$, 3201 (1970).

110. J.J. Gonzales, "Cooperative Reactions on Polymers and Sequential Analysis", Institut for Teoretisk Fysikk, Trondheim, 1977.

111. W.G. Glasser and H.R. Glasser, Macromolecules, $\underline{7}$, 17 (1974).

112. M. Falk and R.E. Thomas, Can. J. Chem., $\underline{52}$ (1974) - cited in ref. 33.

113. M.I. Siling, V.N. Krivsunov et al., Visokomol. Soed., $\underline{16}$, 2550 (1974).

114. P.J. Flory, "Principles of Polymer Chemistry", Cornell Univ. Press, Ithaca, New York, 1953, ch. 9.

115. S.G. Whiteway, I.B. Smith and C.R. Masson, Can. J. Chem., $\underline{48}$, 33 (1970).

116. C.R. Masson, I.B. Smith and S.G. Whiteway, Can.J. Chem. $\underline{40}$, 201, 1456 (1970); $\underline{51}$, 1422 (1973).

117. H. Staudinger, Makromol. Chem., $\underline{1}$, 7 (1947).

118. A.A. Whitehouse, E. Pritchett and G. Barnett, "Phenolic Resins", London, 1967.

119. R. Inoue, T. Minami and T. Ando, $\underline{61}$, 1340 (1958).

120. L.M. Yedanapalli and A.K. Kuriakose, J. Scient. Ind. Res., $\underline{B\ 18}$, 467 (1959).

121. H. Horinti, Gosei Kagaku, $\underline{66}$, 1379 (1963).

122. T. Yoshikawa and I. Kumanotani, Makromol. Chem., $\underline{131}$, 273 (1970)

123. I. Motoc, S. Holban and D. Ciubotariu, Preprint, Timisoara Univ., 1975.

124. R. Vancea, I. Motoc, D. Ciubotariu and S. Holban, Preprint, Timisoara Univ., 1975.

125. P.C. Wu, J.A. Howell and P. Ehrlich, Ind. Eng. Chem. Prod. Res. Develp., $\underline{11}$, 35 (1972).

126. A.M. Kotliar and S. Podgor, J. Polym. Sci., $\underline{55}$, 423 (1961).

127. A.M. Kotliar, J. Polym. Sci., $\underline{A1}$, 3175 (1963).

128. P.J. Meddings and O.E. Potter, Adv. Chem. Series, $\underline{109}$, 96 (1972).

129. J. Malac, J. Polym. Sci., $\underline{C33}$, 233 (1971)

130. J. Malac, J. Polym. Sci., A-1, $\underline{9}$, 3563 (1971).

131. J. Malac, J. Macromol. Sci. - Chem., $\underline{A7}$, 923 (1971)

132. O. Smidsrod and S.G. Whittington, Macromolecules, $\underline{2}$, 42 (1969).

133. E. Tsuchida, K. Mishima, K. Kitamura and I. Shinohara, J. Polym. Sci. Chem., $\underline{10}$, 3615 (1972).

134. E. Tsuchida, T. Yao, K. Kitamura and I. Shinohara, J. Polym. Sci., Chem., $\underline{10}$, 3605 (1972).

135. E. Tsuchida, K. Mishima, K. Kitamura and I. Shinohara, J. Polym. Sci. Chem., $\underline{10}$, 3627 (1972)

136. E. Tsuchida, K. Kitamura and I. Shinohara, J. Polym. Sci. Chem., $\underline{10}$, 3639 (1972).

137. R.H. MacDowell and E. Percival, "Chemistry and Enzymology of Marine Algol Polysaccharides", Academic Press, New York 1967.

138. T. Fueno and J. Furukawa, J. Polym. Sci. A, $\underline{2}$, 3681 (1964).

61

139. Knuth, The art of computer programming, vol. 2, Addison-Wesley, New York, 1971.

140. F. Gantmacher, Matrizenrechnung, VEB Deutscher Verlag der Wissenschaften, Berlin.

Chapter 2

MONTE CARLO CALCULATION OF SEQUENCE DISTRIBUTIONS IN POLYMERS.

I. Motoc and K.F. O'Driscoll

2.1. Introduction

2.2. Short review of analytical models, Monte Carlo algorithms and
       computer programs.

    2.2.1. Copolymerization digraphs

2.3. Irreversible copolymerizations

    2.3.1. Binary copolymerization with ultimate effect

    2.3.2. Binary copolymerization with penultimate effect

    2.3.3. Terpolymerization

2.4. Reversible copolymerizations

2.5. Configurational sequences in stereoregular polymers.

    References

2.1. Introduction

The aim of this chapter is to discuss in detail the Monte Carlo algorithms developed to compute the sequence distributions in polymers. Because stereoregular polymers constitute a unique form of copolymer, the stereosequence distributions in vinyl homopolymers and the sequence distributions in copolymers can be computed using the same algorithms. Also included is a brief review of probabilistic models (i.e., Bernoulli trials and Markov chains) frequently used to compute the sequence distribtuion.

The determination of sequence distributions is important for the understanding of polymer physical properties, to compute the monomer reactivity parameters and to discriminate among polymerization mechanisms.

2.2. Short review of analytical models, Monte Carlo algorithms and computer programs.

A Bernoullian model was developed[1] by Price. Within this model the probability of a given state of the system is independent of the previous state and does not condition the next state. The Bernoullian behaviour has been shown to describe cis-trans distributions among 1, 4 additions in polybutadienes[2-4], the comonomer distribution in ethylene-vinyl acetate copolymer[5], and configurational distributions in polystyrene[6], poly (vinyl chloride)[7], poly (vinyl alcohol)[7]

Consider the binary copolymerization:

$$- M_I^* + M_J \rightarrow - M_I M_J^* \qquad ; \quad I, J = 1, 2 \qquad (1)$$

where $- M_I^*$, $I = 1,2$, is an ionic or radical polymeric chain end, and M_J, $J = 1,2$, is a monomer. Because the final state (i.e., $- M_J^*$) of the system (1) may be regarded as independent of its initial state (i.e., $- M_I^*$), the final state $- M_1^*$ occurs with the unconditional probability P_1 and $-M_2^*$ with probability P_2. Obviously:

$$P_1 + P_2 = 1 \qquad (2)$$

The sequence distribution $n_I(m)$, where I denotes the chemical identity of the monomeric unit of the homogeneous sequence and m the sequence length, is given by:

$$n_I(m) = (P_I)^m \ , \ I = 1,2 \ \text{(mole fraction)} \tag{3}$$

For example, the relative concentrations of the sequences M_1M_1, $M_1M_1M_1$, M_2M_2 etc. are $P_1{}^2$, $P_1{}^3$, $P_2{}^2 = (1 - P_1)^2$ etc. The mole fractions of heterogeneous sequences are computed similarly. For M_1M_2, $M_1M_2M_1$, $M_1M_2M_2M_1$, etc. one obtains $P_1P_2 = P_1(1 - P_1)$, $P_1P_2P_1 = P_1{}^2 (1 - P_1)$, $P_1P_2P_2P_1 = P_1{}^2 (1 - P_1)^2$ etc. P_1 and $P_2 = (1 - P_1)$ correspond to the M_1 and M_2 mole fraction observed for the polymer.

In the case of Bernoullian behavior, the number-average sequence lengths:

$$\bar{n}_I = \sum_m m \, n_I(m) \ / \sum_m n_I(m) \quad , \ I = 1,2 \tag{4}$$

are:

$$\bar{n}_1 = \frac{1}{1 - P_1} \quad \text{and} \quad \bar{n}_2 = \frac{1}{P_1} \tag{5}$$

(for details, one may consult ref. 9). The above equations are easily generalized for multicomponent polymerizations.

Markov chain models were developed by Coleman[12], Price[13], Hijmans[14] and others (for a review, see ref. 11). These models take into account those situations where a given state of the system is dependent upon the previous states. Most Markov chain models applications deal with first-order Markov chains (i.e., the given state depends upon a single preceding state).

For binary irreversible copolymerizations with ultimate effect (1) the transition probabilities from the initial state - M_I^* to the final state - M_J^* are denoted by P_{IJ}, I, J = 1,2. The sequence distribution is computed according to equation

$$n_1(m) = P_{21} \, P_{11}^{m-1} \, P_{12} \quad \text{and} \quad n_2(m) = P_{12} \, P_{22}^{m-1} \, P_{21} \tag{6}$$

Because among P_{IJ} values the following relations hold

$$P_{11} + P_{12} = P_{22} + P_{21} = 1 \tag{7}$$

equations (6) may be rewritten as:

$$n_1(m) = P_{21} \, P_{11}^{m-1} (1 - P_{11}) \quad \text{and} \quad n_2(m) = P_{12} \, P_{22}^{m-1} (1 - P_{22}) \tag{8}$$

The number-average sequence lengths are given by equations:

$$\bar{n}_1 = 1/P_{12} \quad \text{and} \quad \bar{n}_2 = 1/P_{21} \tag{9}$$

$P_{IJ}/(P_{IJ} + P_{JI})$, $I \neq J$, corresponds to the mole fraction of the monomer M_J in copolymer.

Rudin et al. have shown recently[15] that $\bar{n}_I$, $I = 1,2$, may be calculated from a combination of copolymer and triad compositions:

$$\bar{n}_1 = (f_{212} + \tfrac{1}{2} f_{112})^{-1} \quad \text{and} \quad \bar{n}_2 = (m_2/m_1)\bar{n}_1 \tag{10}$$

where m_I is the mole fraction of monomer M_I in the copolymer and f_{IJK}, I, J, $K=1,2$, are the triad mole fractions normalized according to relation

$$f_{111} + f_{212} + f_{112} = 1$$

It is interesting to note that equations (9) and (10) allow the estimation of the reactivity ratios from sequence distribution, namely:

$$\bar{n}_1 = \frac{1}{P_{12}} = (f_{212} + \tfrac{1}{2} f_{112})^{-1} = r_1 \frac{[M_1]}{[M_2]} + 1 \tag{11}$$

$$\bar{n}_2 = \frac{1}{P_{21}} = (m_2/m_1)\bar{n}_1 = r_2 \frac{[M_2]}{[M_1]} + 1 \tag{12}$$

where $[M_I]$, $I = 1,2$, is the concentration of monomer I in the feed. Equations (11) and (12) were applied[16] by O'Driscoll to a study of styrene - vinyl acetate copolymers.

For binary irreversible copolymerizations with penultimate effect:

$$- M_I M_J{}^* + M_K \rightarrow - M_J M_K{}^* \quad , \quad I, J, K = 1, 2 \tag{13}$$

the transition probabilities are denoted by P_{IJK}. For example, P_{121} signifies the probability for transition from the initial state $- M_1 M_2{}^*$ to the final state $- M_2 M_1{}^*$. The sequence distribution is given by:

$$n_I(m) = P_{JII} \, P_{III}^{m-2} \, P_{IIJ} \quad , \quad I, J = 1, 2 \text{ and } I \neq J \tag{14}$$

Because relations (15) take place

$$P_{IJ1} + P_{IJ2} = 1, \quad I, J = 1, 2 \tag{15}$$

equation (14) is equivalent to equation

$$n_I(m) = P_{JII} \, P_{III}^{m-2} \, (1 - P_{III}) \tag{16}$$

The number-average sequence lengths are computed as:

$$\bar{n}_1 = 1 + P_{211}/P_{112} \text{ and } \bar{n}_2 = 1 + P_{122}/P_{221} \tag{17}$$

(for technical details one may consult ref. 10).

For irreversible terpolymerization with ultimate effect

$$- M_I^* + M_J \rightarrow - M_J^* , I, J = 1, 2, 3 \tag{18}$$

one defines the probabilities P_{IJ} for the transition from the initial state $- M_I^*$ to the final state $- M_J^*$, I, J = 1, 2, 3.

For a given I, the P_{IJ} values are normalized, i.e.,

$$\sum_{j=1}^{3} P_{IJ} = 1, \quad I = 1, 2 \text{ or } 3 \tag{19}$$

The sequence distribution and the number-average sequence lengths are given by equations (20) and (21), respectively:

$$n_I(m) = P_{II}^{m-1} (P_{IJ} + P_{IK}) , \text{ and} \tag{20}$$

$$\bar{n}_I = 1/(P_{IJ} + P_{IQ}) , I, J, Q = 1, 2, 3 \text{ and } I \neq J, I \neq Q \tag{21}$$

Computer programs which implement the above models for binary irreversible copolymerizations (ultimate and penultimate effect) and irreversible terpolymerizations (ultimate effect) are available in refs. 18 and 19, respectively.

The calculation of the sequence distribution of like configurations and the number-average sequence length of like, meso and racemic seuqences in stereo-regular homopolymers conforms formally[9] to the equations established for copolymers.

The Monte Carlo simulation represents a useful (and often the only alternative) method to perform polymer structure calculations. The general features of the Monte Carlo models are described in the previous chapter. The present chapter reviews in detail the algorithms developed to compute the sequence distribution in polymers. The advantages of the Monte Carlo models are: i) they may easily take into account the reaction mechanism; ii) they provide (approximate) informations on polymer structure for small as well as high conversions; and iii) they offer the possibility to visualize some features of the polymer chain. The main disadvantage of Monte Carlo methods consists in relatively large amounts of computer time spent to obtain statistically reliable results.

2.2.1. Copolymerizations digraphs.

One may attach to a copolymerization reaction a directed graph (digraph) which describes the interrelations among elementary steps of the reaction. These digraphs are useful as "flow charts" for Monte Carlo algorithms.

A digraph G is defined[20] by the doublet $G = (X,\Gamma)$. X is a finite set whose elements are termed vertices and Γ is an application of X onto X, i.e., $\Gamma: X \to X$. If $\Gamma(x_1) = x_2$, x_1, $x_2 \in X$, the vertices x_1 and x_2 are connected by an arc:

$$x_1 \qquad x_2$$

$\Gamma(x_1) = x_1$ defines a loop:

and, if $\Gamma(x_1) = \phi$ (i.e., the empty set), the vertex x_1 is disconnected,

Consider[19] the copolymerization reactions (1), (13) and (18).

If one labels the active species involved in copolymerizations (1) as:

$$- M_1{}^* \leftrightarrow 1 \qquad \text{and} \qquad - M_2{}^* \longleftrightarrow 2$$

it is easily seen that the elementary steps may be interrelated as:

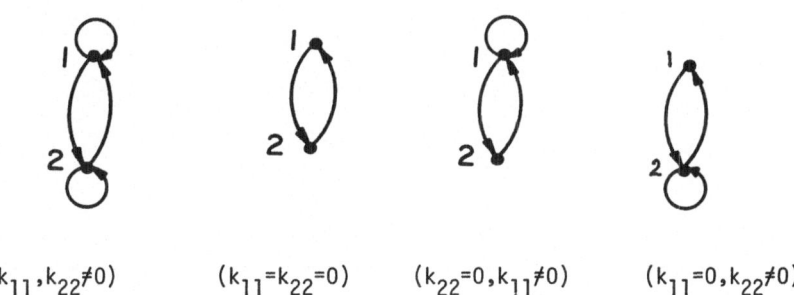

$$(k_{11}, k_{22} \neq 0) \qquad (k_{11} = k_{22} = 0) \qquad (k_{22} = 0, k_{11} \neq 0) \qquad (k_{11} = 0, k_{22} \neq 0)$$

The complete digraph for copolymerizations with penultimate effect (13) is:

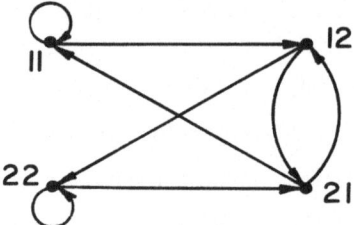

and for terpolymerizations with ultimate effect (18) the complete digraph is:

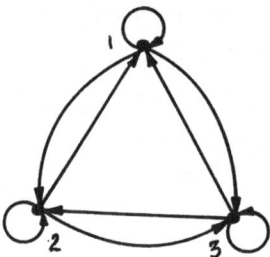

(where I = 1, 2, 3 stands for $- M_I*$)

($k_{II} \neq 0$)

2.3. Irreversible Copolymerizations.

The Monte Carlo algorithms developed for irreversible copolymerizations are systematized below.

Irreversible copolymerizations[a]: Monte Carlo algorithms

Copolymerizations	Authors	Ref.
1. Binary (ultimate effect)	Smidstrod and Wittington	21
	Marconi et al.	22,23
	Saito and Matsumura	24
	Motoc et al.	25-31
	O'Driscoll	32
	Mirabella	33
2. Binary (penultimate effect)	Motoc et al.	34-36
3. Ternary (ultimate effect)	Mirabella	37
	Motoc et al.	38

[a] Computer programs are available in refs. 28,33,37 and in chapter 4 of this book.

2.3.1. Binary copolymerization with ultimate effect.

A binary copolymerization with ultimate effect (i.e., only the ultimate monomeric unit affects the rate constants) involves[39,40] four propagation reactions:

$$- M_I^* + M_J \xrightarrow{\quad k_{IJ} \quad} - M_J^* \quad , \quad I, J = 1,2$$

The ratios $r_I = k_{II}/k_{IJ}$, I, J = 1,2, I≠ J are termed reactivity ratios. Assuming that i) the rate constants are independent of the length of the growing chain, and ii) one may neglect the initiation - termination processes (i.e., one considers the polymers whose length is so great that influences of initiation and termination processes are absent), the transition probabilities P_{IJ} from the state - M_I^* to the state - M_J^* are easily computed:

$$P_{II} = \frac{k_{II}[-M_I^*][M_I]}{k_{II}[-M_I^*][M_I] + k_{IJ}[-M_I^*][M_J]} =$$

$$= \frac{k_{II}[M_I]}{k_{II}[M_I] + k_{IJ}[M_J]} = \frac{r_I[M_I]}{r_I[M_I] + [M_J]} =$$

$$= \frac{r_I}{r_I + C} \quad , \quad I = 1,2 \quad , \quad I \neq J \tag{22}$$

where [] denotes the feed concentration and $C = [M_J]/[M_I]$. P_{IJ} is obtained from the relation (7):

$$P_{IJ} = 1 - P_{II} = \frac{C}{r_I + C} \tag{23}$$

The Monte Carlo simulation algorithm of propagation reactions consists of the following steps:

1) specify the first monomeric unit of the chain, say M_I, C, r_1 and r_2 values and the length of the chain to be formed (degree of polymerization), N.

2) Generate the random number $\xi \in (0,1)$, ξ's being uniformly repartised within this interval.

3) Test the inequality:

$$\xi < P_{II} \tag{24}$$

If it holds, the mer M_I is added, the system remains in the state - M_I^* and the step (2) is resumed. If (24) does not hold, go to step (4).

4) The mer M_J, $J \neq I$ is added, the system passes into the state - M_J^*, the index I becomes J and the step (2) is resumed.

5) The computing procedure is stopped when the desired polymerization degree is attained.

The flow chart of our MEMORY-3 program (see chapter 4) is shown in Figure 1.

The input data are: the first monomeric unit of the chain, reactivity ratios r_1 and r_2, monomer feed concentrations $[M_1]$ and $[M_2]$, polymerization degree N and the variable FRMOL = 1 or 2. When FRMOL = 1 the simulation is performed under stationary conditions, i.e., $[M_1]$, $[M_2]$ = constant. FRMOL = 2 implies to reevaluate the feed concentrations after each monomeric unit addition. One achieves this aim as follows: $C_{(0)} = [M_J]_{(0)}/[M_I]_{(0)}$ is interpreted as the ratio of the number of monomers M_J and M_I, respectively, per growing specie at the beginning of simulation. Introducing the δ-type functions

$$\delta_I = \begin{array}{l} 0, \text{ if the added monomer in the previous addition step is } M_J \\ 1, \text{ if the added monomer in the previous addition step is } M_I \end{array} \qquad (25)$$

The value of C-ratio in the n addition step is:

$$C_{(n)} = \frac{[M_J]_{(n-2)} - \delta_J}{[M_I]_{(n-2)} - \delta_I} = \frac{[M_J]_{(n-1)}}{[M_I]_{(n-1)}} \qquad , \quad n \geq 1 \qquad (26)$$

Introducing $C_{(n)}$ defined by relation (24) into equations (22) and (23) the transition probabilities become a function of conversion.

MEMORY-3 program uses the best sequence of a thousand random numbers obtained by means of ALEAT generator, i.e., the sequence 6000-7000 (we proved in chapter 1 that a polymerization degree $N \geq 500$ assures an acceptable statistical stability). In this way, one preserves the important advantage to visualize the features of the polymer chain.

The output of MEMORY-3 is a report of input parameters, polymer composition, the statistic of homogeneous sequences $n_I(m)$, I = 1, 2 and m_i = 1, 2,..., and the relative position of monomeric units within the chain of polymerization degree N.

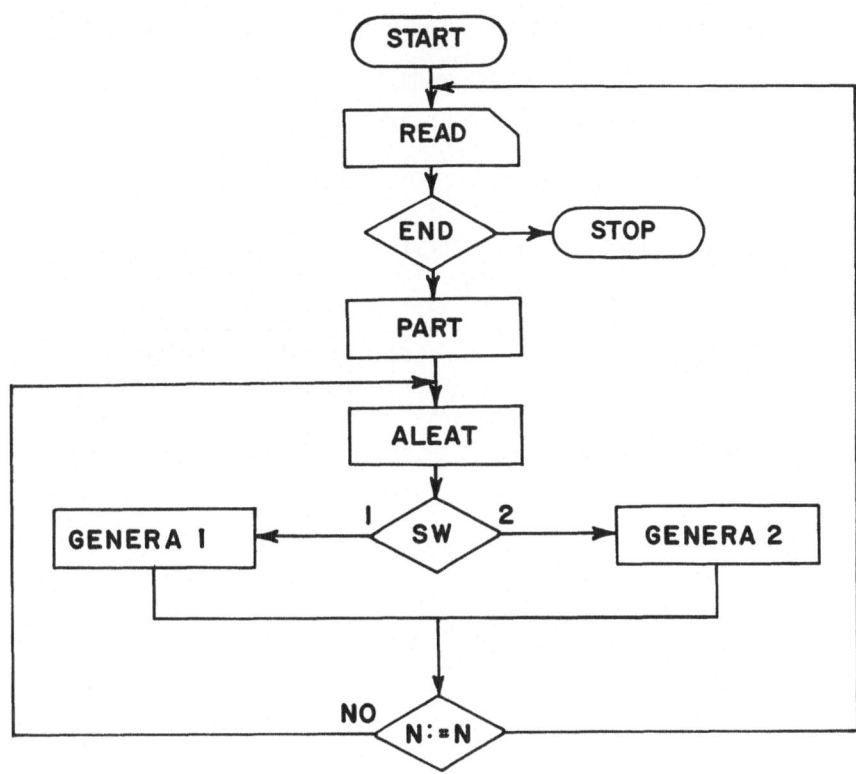

Figure 1. The flow chart of MEMORY-3 program.

Illustratively, for the copolymerization of acrylic acid (M_1) and methyl acrylate (M_2) the reactivity ratios[41] are r_1 = 1.4 and r_2 = 1.0. Tables 1-4 collect the results obtained by Monte Carlo simulation[36] and using the first order Markov chain model (6).

Table 1. Composition of acrylic acid (m_1)/methyl acrylate (M_2) copolymers.

	M_1 in feed (mole %)	M_1 in copolymer (mole %)		
		Experimental	Markov (M)	Monte Carlo (MC)
1.	88.0	91.1	90.8	92.0
2.	81.0	85.7	84.9	85.4
3.	72.4	77.2	77.2	76.7
4.	71.6	75.6	76.4	75.5
5.	58.9	61.5	63.9	61.2
6.	58.7	62.2	63.7	61.1
7.	51.0	56.1	55.6	53.4
8.	45.8	49.1	50.0	46.6
9.	34.0	41.1	36.9	33.0
10.	30.5	32.6	33.0	28.7
11.	20.2	25.1	21.5	20.2

The agreement between experimental and computed compositions is very good, as argued by the following least-squares equations:

$$\%M_{1,exp} = 7.658 + 0.904\ \%M_{1,MC}$$

$$(r = 0.998,\ s = 1.371,\ F = 981.779)$$

$$\%M_{1,exp} = 2.854 + 0.957\ \%M_{1,M}$$

$$(r = 0.996,\ s = 1.759,\ F = 594.962)$$

where r is the correlation coefficient, s is the standard deviation and F is Fisher statistics.

Table 2. Sequence distribution in acrylic acid/methyl
acrylate copolymer no. 1 (Table 1)

m	Monte Carlo $n_1(m)$	$n_2(m)$	Markov $n_1(m)$	$n_2(m)$
1	3	72	7.15	70.95
2	5	4	6.52	8.51
3	6		5.94	1.02
4	8		5.41	
5	3		4.93	
6	5		4.49	
7	4		4.09	
8	4		3.73	
9	1		3.40	
10	3		3.10	
11	2		2.82	
12	1		2.57	
13	2		2.34	
14	-		2.14	
15	1		1.95	
16	5		1.77	
17	6		1.62	
18	1		1.47	
19	1		1.34	
20	1		1.22	
21	-		1.11	
22	3		1.01	
23	2			
24	2			
25	-			
26	3			

Table 2 (continued)

27	1
28	1
34;38	2;1

Table 3. Sequence distribution in acrylic acid (M_1)/methyl
acrylate (M_2) copolymer no. 2.

m	Monte Carlo		Markov	
	$n_1(m)$	$n_2(m)$	$n_1(m)$	$n_2(m)$
1	15	103	17.49	98.74
2	11	20	14.98	18.76
3	14	1	12.83	3.56
4	13		10.99	
5	6		9.41	
6	15		8.06	
7	6		6.90	
8	7		5.91	
9	3		5.06	
10	5		4.34	
11	5		3.72	
12	6		3.18	
13	5		2.73	
14	2		2.33	
15	5		2.00	
16	2		1.71	
17	1		1.47	
18	-		1.25	
19	1		1.08	

Table 4. Sequence distribution in acrylic acid (M_1)/methyl acrylate (M_2) copolymer no. 3.

m	Monte Carlo		Markov	
	$n_1(m)$	$n_2(m)$	$n_1(m)$	$n_2(m)$
1	43	133	35.35	119.59
2	27	34	27.78	33.01
3	24	8	21.84	9.11
4	16	2	17.16	2.51
5	18		13.49	
6	15		10.60	
7	8		8.33	
8	7		6.55	
9	5		5.15	
10	4		4.05	
11	4		3.18	
12	-		2.50	
13	2		1.96	
14	-		1.54	
15	3		1.21	

Inspecting Tables 2, 3 and 4 one concludes that statements concerning the sequence distribution are fragile[42], i.e., they strongly depend on the details of the model used.

For the copolymerization of methyl methacrylate (M_1) with chloroprene (M_2) Braun et al. proposed[43] an ultimate effect mechanism $(r_1 = 0.08$ and $r_2 = 5.1)$, while Ebdon[44] a semi-penultimate effect mechansim $(r_{11} = 0.107, r_{21} = 0.057$ and $r = 6.7)$. Our MEMORY-3 type results[30b] argue for an ultimate effect mechanism. The data are collected in Tables 5 and 6.

Moreover, one favours the ultimate effects because it uses two adjustable parameters (i.e., r_1, r_2) instead of three (i.e., r_{12}, r_{21}, r_2) in the case of the semi-penultimate effect.

Table 5. Methyl methacrylate (M_1)/chloroprene (M_2)
copolymers: composition.

No.	M_1 in feed, mole-fraction	M_1 in copolymer, mole-fraction		
		Experimental	Monte Carlo[30b]	Ebdon[44]
1.	0.98	0.86 ± 0.02	0.82	0.84
2.	0.97	0.77 ± 0.02	0.75	0.78
3.	0.96	0.71 ± 0.02	0.70	0.72
4.	0.95	0.64 ± 0.02	0.65	0.67
5.	0.93	0.56 ± 0.02	0.59	--
6.	0.90	0.48 ± 0.02	0.51	0.50
7.	0.85	0.40 ± 0.02	0.42	--
8.	0.80	0.34 ± 0.02	0.36	0.33
9.	0.70	0.22 ± 0.02	0.24	0.23
10.	0.60	0.18 ± 0.02	0.18	0.17
11.	0.50	0.13 ± 0.02	0.14	0.12
12.	0.40	0.11 ± 0.01	0.10	--
13.	0.30	0.08 ± 0.01	--	--

Table 6. Methyl methacrylate (M_1)/chloroprene (M_2) copolymers:
$M_1M_1M_1$ and $M_2M_1M_2$ triad distribution[a])

Copolymer no.	Experimental		Monte Carlo[30b]		Ebdon[44]	
	%$M_1M_1M_1$	%$M_2M_1M_2$	%$M_1M_1M_1$	%$M_2M_1M_2$	%$M_1M_1M_1$	%$M_2M_1M_2$
1.	92.3	7.7	92.0	8.0	94.7	5.3
2.	84.1	15.9	87.5	12.5	88.2	11.8
3.	70.6	29.4	78.7	21.2	80.9	19.5
4.	64.9	35.1	66.2	33.8	71.7	28.3
6.	36.2	63.8	29.5	70.5	35.6	64.4
8.	8.8	91.2	4.9	95.1	8.6	91.4

[a]) Copolymer no. corresponds to Table 5.

The program developed[32] by O'Driscoll computes the pentad distribution and has been used to test reactivity ratios values derived from NMR spectra[45]. A sample of typical result is given below ($r_1 = 0.20$, $r_2 = 0.20$; feed mole fraction of monomer M_1 is 0.50).

The first 100 monomeric units of the copolymer with polymerization degree 10,000 are:

$M_1M_2M_1M_2M_2M_1M_2M_1M_2M_1M_2M_1M_1M_2M_2M_1M_2M_2M_1M_2M_1M_2M_2M_2M_1M_2M_1M_2M_1M_2M_1M_2M_1M_1M_2M_1M_2M_1M_2M_1$

$M_2M_2M_1M_2M_1M_2M_2M_1M_2M_2M_1M_2M_1M_2M_2M_1M_2M_1M_2M_1M_2M_1M_1M_2M_2M_1M_2M_1M_2M_1M_2M_1M_2M_1M_2M_2M_2M_1M_2M_2M_1M_2$

$M_1M_1M_2M_1M_2M_1M_2M_1M_2M_1M_2M_1M_2M_2M_2M_1M_2M_2M_1M_1$

The fractions of M_I, $I = 1, 2$, centered pentads are systhematized below:

Pentad	Fraction of M_1 centered pentads	Fraction of M_2 centered pentads
$M_2M_2M_1M_2M_2$	0.024	
$M_2M_2M_1M_2M_1$	0.205	
$M_2M_2M_1M_1M_2$	0.046	
$M_2M_2M_1M_1M_1$	0.008	
$M_2M_1M_1M_2M_1$	0.200	
$M_2M_1M_1M_1M_2$	0.017	
$M_1M_2M_1M_2M_1$	0.455	
$M_2M_1M_1M_1M_1$	0.009	
$M_1M_2M_1M_1M_1$	0.035	
$M_1M_1M_1M_1M_1$	0.001	
$M_1M_1M_2M_1M_1$		0.021
$M_1M_1M_2M_1M_2$		0.189
$M_1M_1M_2M_2M_1$		0.047
$M_1M_1M_2M_2M_2$		0.007
$M_1M_2M_2M_1M_2$		0.206
$M_1M_2M_2M_2M_1$		0.020
$M_2M_1M_2M_1M_2$		0.455

(continued)

Pentad	Fraction of M_1 centered pentads	Fraction of M_2 centered pentads
$M_1M_2M_2M_2M_2$		0.010
$M_2M_1M_2M_2M_2$		0.044
$M_2M_2M_2M_2M_2$		0.001

Total number of pentads: 9996

Total number of M_1 centered pentads: 4966

Total number of M_2 centered pentads: 5030

The program developed[33] by Mirabella takes into account also the initiation step:

$$I\cdot + M_1 \longrightarrow I - M_1\cdot$$

$$I\cdot + M_2 \longrightarrow I - M_2\cdot$$

The initiation reactions rate constants, k_{i1} and k_{i2}, are rarely known. Accordingly, the assumption that the initiator selects between monomers M_1 and M_2 on the bases of feed composition, $[M_1]$ and $[M_2]$, and relative monomer reactivities, I_1 and I_2, toward the same radical $I\cdot$ was made. Thus, the probabilities of initiation for $I - M_1$ and $I - M_2$ chains, respectively, are:

$$P_{i1} = \frac{I_1[M_1]}{I_1[M_1] + I_2[M_2]} \quad , \quad P_{i2} = \frac{I_2[M_2]}{I_1[M_1] + I_2[M_2]}$$

So, the first monomeric unit of the chain is M_1 if

$$\xi < P_{i1}$$

otherwise it is M_2. ξ's are random numbers uniform distributed within the interval (0,1). The propagation steps are simulated according to the algorithm described above. From this study it was concluded that Monte Carlo simulation of polymerization permits the realistic prediction of the cumulative copolymer composition as a function of polymerization degree (see Figures 5-7 of ref. 33).

2.3.2. <u>Binary copolymerization with penultimate effect.</u>

The penultimate effect model of vinyl copolymerization introduced[50] by Merz et al., proposes that the penultimate unit, as well as the terminal unit, can influence the addition of the next monomer unit to the growing chain. Thus, it is possible that eight propagation reactions and four reactivity ratios can contribute to the propagation path:

$$- M_I M_J^* + M_K \longrightarrow - M_J M_K^* \quad , \quad I, J, K = 1,2$$

and

$$r_1 = \frac{k_{111}}{k_{112}} \quad , \quad r_1' = \frac{k_{211}}{k_{212}} \quad , \quad r_2 = \frac{k_{222}}{k_{221}} \quad , \quad r_2' = \frac{k_{122}}{k_{121}}$$

The penultimate effect model has been extensively applied to explain composition/conversion and sequence distribution data which did not fit the ultimate effect model[46-48].

The Monte Carlo algorithm to simulate the binary copolymerization with penultimate effect has been developed[34] by Motoc et al.

The transition probabilities from the state $- M_I M_J^*$ to the state $- M_J M_K^*$, i.e., P_{IJK}, are computed according to the equation:

$$P_{IJK} = \frac{k_{IJK}[-M_I M_J^*][M_K]}{k_{IJK}[-M_I M_J^*][M_K] + k_{IJL}[-M_I M_J^*][M_L]} = \qquad (27)$$

$$= \frac{k_{IJK}[M_K]}{k_{IJK}[M_K] + k_{IJL}[M_L]} \quad ; \quad I, J, K, L = 1, 2, \quad K \neq L$$

The eight P_{IJK} values are:

$$P_{111} = \frac{r_1}{c + r_1} \qquad , \qquad P_{112} = \frac{c}{c + r_1}$$

$$P_{222} = \frac{r_2}{c^{-1} + r_2} \qquad , \qquad P_{221} = \frac{c^{-1}}{c^{-1} + r_2}$$

$$P_{122} = \frac{r_2'}{c^{-1} + r_2'} \qquad , \qquad P_{121} = \frac{c^{-1}}{c^{-1} + r_2'} \qquad (28)$$

$$P_{211} = \frac{r_1'}{c + r_1'} \qquad , \qquad P_{212} = \frac{c}{c + r_1'} \quad , \text{ where}$$
$$c = [M_2]/[M_1].$$

It is obvious that:

$$0 \leq P_{IJK} \leq 1 \quad \text{and} \quad P_{IJK} + P_{IJL} = 1, \quad I, J, K, L = 1, 2, \quad K \neq L$$

The algorithm implemented by the MEMORY-5 program (see chapter 4) consists of the following steps:

1) specify the first two monomeric units of the chain, say $M_I M_J$, C, r_1, r_1', r_2 and r_2' values and the degree of polymerization, N, the value of FRMOL (see MEMORY-3).

2) generate the random number $\xi \, \epsilon (0,1)$, ξ's being uniformly repartised within this interval.

3) test the inequality:

$$\xi < P_{IJ1}$$

If it holds, the mer M_1 is added, the system passes in the state - $M_J M_1$ and step (2) is resumed. If the inequality does not hold, go to step (4).

4) the mer M_2 is added, the system passes into the state - $M_J M_2$ and the step (2) is resumed.

5) The computing procedure is stopped when the desired polymerization degree is attained.

One may easily take into account the conversion via δ function (25), i.e., introducing $C_{(n)}$ defined by equation (26) into equations (28).

The flow chart of MEMORY-5 program is shown in Figure 2 (the complete listing of the program is given in chapter 4).

The output of MEMORY-5 is a report of input parameters, copolymer composition, the statistic of homogenous sequences $n_I(m)$, I = 1, 2, and m = k, 2, ..., and the relative position of monomeric units within the chain of polymerization degree N.

The Monte Carlo study[35] of copolymerization of styrene with four esters of benzylidenecyanoacetic acid furnished some interesting results. The feed and copolymer compositions, r_1, r_2 and r_1, r_1', r_2, r_2' values are taken from ref. 49. The reactivity ratios used in calculations are collected in Table 7.

Table 8 displyes the copolymers composition computed by means of MEMORY-3

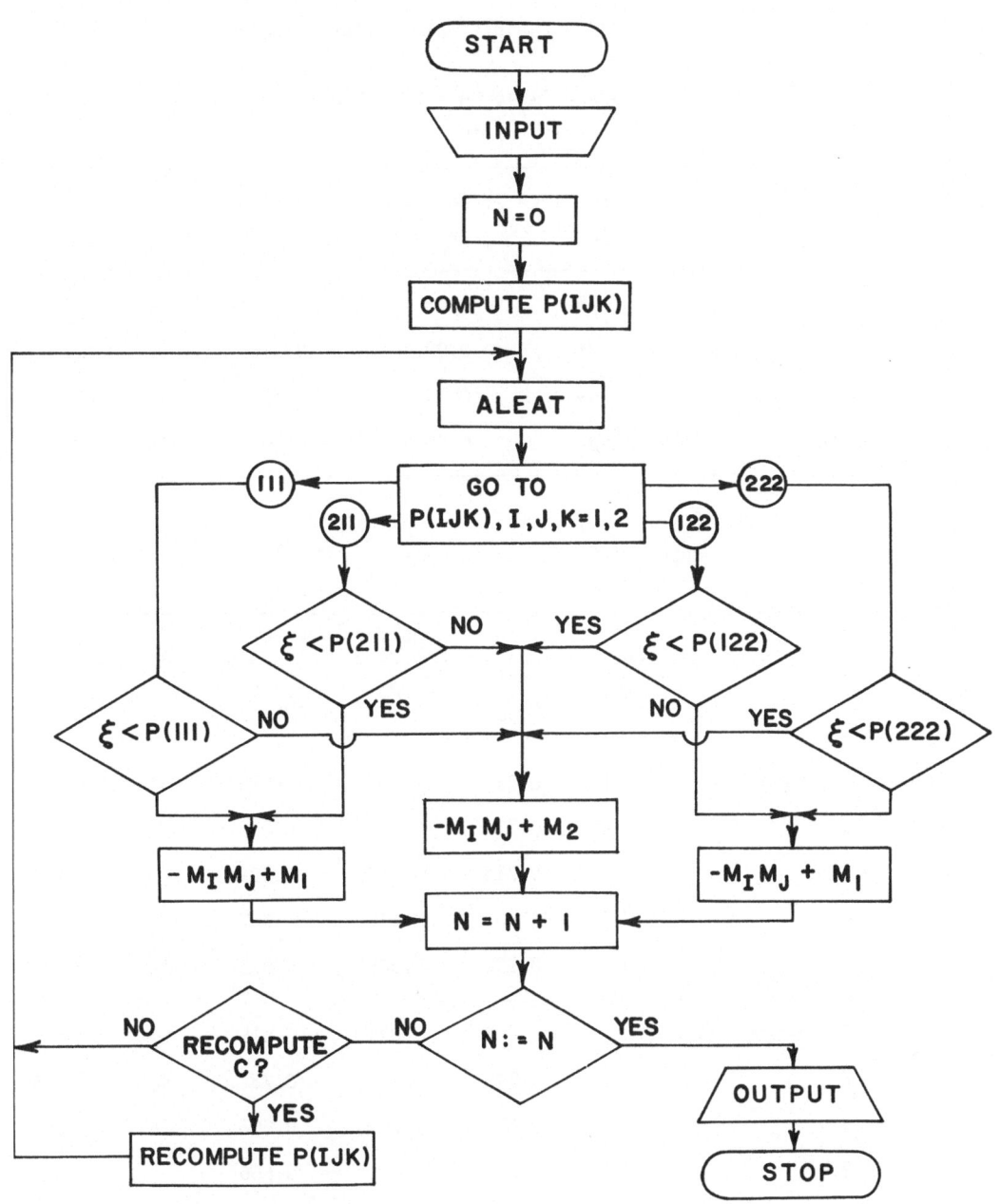

Figure 2. The flow chart of MEMORY-5 program.

(ultimate effect) and MEMORY-5 (penultimate effect) program.

Table 7. Styrene $(M_1)\Phi$ CH $=$ C(CN)COOR (M_2) copolymers:
reactivity ratios

R	Ultimate effect		Penultimate effect		
	r_1	r_2	r_1	r_1'	$r_2=r_2'$
Methyl	0.45	0.00	0.325	1.330	0.00
n-Hexyl	0.42	0.00	0.315	1.000	0.00
Cyclohexyl	0.44	0.00	0.300	0.925	0.00
Benzyl	0.43	0.00	0.340	0.715	0.00

Table 8. Styrene (M_1)/esters of benzylidenecyanoacetic acid (M_2)
copolymers: composition.

No.	R	M_1 in feed, mole-fraction	M_1 in Experimental $(M_{1,E})$	copolymer, Memory-3 $(M_{1,M3})$	mole-fraction M ory-5 $(M_{1,M5})$	Δ_1^2 $(\times 10^{-6})$	Δ_2^2 $(\times 10^{-6})$
1.	Methyl	0.9	0.821	0.833	0.818	144	9
2.		0.8	0.729	0.721	0.745	64	256
3.		0.7	0.715	0.656	0.699	3481	256
4.		0.6	0.652	0.617	0.667	1225	225
5.		0.5	0.597	0.586	0.638	121	1681
6.	Hexyl	0.9	0.815	0.821	0.818	36	9
7.		0.8	0.745	0.711	0.737	1156	64
8.		0.7	0.677	0.648	0.693	841	256
9.		0.6	0.613	0.611	0.656	4	1849
10.		0.5	0.637	0.583	0.630	2916	49

Table 8 (continued)

11.	Cyclohexyl	0.9	0.808	0.829	0.812	441	16
12.		0.8	0.742	0.720	0.735	484	49
13.		0.7	0.685	0.653	0.689	1024	16
14.		0.6	0.644	0.616	0.652	784	64
15.		0.5	0.631	0.585	0.627	2116	16
16.	Benzyl	0.9	0.820	0.827	0.814	49	36
17.		0.8	0.729	0.715	0.732	196	9
18.		0.7	0.692	0.651	0.678	1681	196
19.		0.6	0.634	0.613	0.637	441	9
20.		0.5	0.610	0.584	0.607	676	9

The Δ values in Table 8 represent:

$$\Delta_1^2 = (M_{1,E} - M_{1,M3})^2 \quad \text{and} \quad \Delta_2^2 = (M_{1,E} - M_{1,M5})^2$$

The standard deviation, s, of copolymer composition from the ultimate, respective penultimate effect model are computed according to the relation:

$$s_{M3} = [\frac{1}{n-1} \sum_{i=1}^{n} \Delta_{1,i}^2]^{1/2} \quad \text{and} \quad s_{M5} = [\frac{1}{n-1} \sum_{i=1}^{n} \Delta_{2,i}^2]^{1/2}$$

One can prove[51] that the ratio s_{M3}^2 / s_{M5}^2 , where $s_{M3}^2 > s_{M5}^2$ is repartised as Fisher F-values. Thus, one may use the F-test in order to conclude if the ultimate or penultimate model is obeyed. The results are recorded in Table 9. Thus, for the methyl and n-hexyl esters the composition data do not supply the necessary information to reject the ultimate effect (two ajustable parameters) in favour of penultimate effect (four ajustable parameters). These results are in good agreement with the statistics (i.e., correlation coefficient, r, standard deviation, s, Fisher statistic, F, and explained variance, EV) corresponding to equation:

$$M_{1,E} = a + b \, M_{1,M3} \text{ (or M5)} \tag{29}$$

Table 9. Utimate vs penultimate effect: the F-values

R	F	Observation	
Methyl	2.07	ultimate effect	$(F < F_{crt})$
n-Hexyl	2.22	ultimate effect	$(F < F_{crt})$
Cyclohexyl	30.76	penultimate effect	$(F > F_{crt})$
Benzyl	11.89	penultimate effect	$(F > F_{crt})$

(F_{crt} values are $F_{0.95,4,4}$ = 6.39 and $F_{0.99,4,4}$ = 15.98)

The sequence distribution data collected in Table 11 argue that one may easily discriminate among the two models using this type of information.

Table 10. $M_{1,E}$ vs $M_{1,M3}$ (I) and $M_{1,E}$ vs $M_{1,M5}$ (II) correlations

R		a	b	r	s	F	EV
Methyl	(I)	0.8328	0.1343	0.966	0.022	13.88	0.910
	(II)	1.1660	-0.1287	0.976	0.018	19.90	0.936
n-Hexyl	(I)	0.8466	0.1261	0.971	0.019	16.55	0.924
	(II)	1.0806	-0.0666	0.969	0.020	15.62	0.919
Cyclohexyl	(II)	0.9975	0.0007	0.996	0.006	137.11	0.996
Benzyl	(II)	1.0111	-0.0043	0.996	0.007	139.06	0.996

Table 11. Computed sequence distribution in copolymers no. 2 and 7 from Table 8. (all M_2 sequences are of the type $(-M_2-)_1$).

m	$n_1(m)$ MEMORY-3		MEMORY-5	
	Methyl	n-Hexyl	Methyl	n-Hexyl
2	69	74	99	98
3	47	53	51	53
4	24	22	28	27
5	14	14	17	14
6	10	10	6	7
7	4	5	5	5
8	4	1	5	5
9	3	2	-	-
10	1	1	2	2
11	-	1	-	1
12	-	-	1	1
13	-	-	1	-
14	-	-	-	-
15	1	-	-	-

For the copolymerization of styrene (M_1) with benzylidene malononitrile (M_2) two sets of reactivity ratios were reported: $r_1 = 0.125$, $r_1' = 1.250$, $r_2 = r_2' = 0.000$ (Kressel et al.[52]) and $r_1 = 1.00$, $r_1' = 1.44$, $r_2 = r_2' = 0.00$ (Borrows et al[53]). The experimental and computed composition data are systhematised in Table 12.

Table 12. Composition of Styrene (M_1)/Benzylidene Malononitrile (M_2) Copolymers.

No.	M_1 in feed, (mole %)	M_1 in copolymer (mole %)		Experimental[c]
		Monte Carlo		
		a	b	
1	92.9	77.8	--	76.9
2	89.4	74.5	73.1	74.5
3	85.8	72.2	71.1	72.5
4	81.5	69.9	69.4	69.5
5	69.1	66.3	66.2	66.7
6	59.8	--	64.3	63.3
7	34.3	59.7	60.0	58.7

a - reactivity ratios from ref. 52; b - reactivity ratios from ref. 53; data taken from ref. 52.

Data collected in Table 12, as well as the least squares equations (30) and (31), argue that one cannot discriminate between the two sets of reactivity ratios using only composition data:

$$\% M_{1,exp} = 1.158 + 1.013\ \% M_{1,a} \tag{30}$$

$$(r = 0.996,\ s = 0.593,\ F = 179.745,\ EV = 0.989)$$

$$\% M_{1,exp} = -14.649 + 1.220\ \% M_{1,b} \tag{31}$$

$$(r = 0.997,\ s = 0.457,\ F = 247.580,\ EV = 0.998)$$

2.3.3. Terpolymerization

There are nine propagation reactions[54,55] which need to be considered in order to describe the terpolymerization propagation steps assuming the ultimate effect model:

$$- M_I^* + M_J \longrightarrow - M_J^*$$

These nine progagation reactions lead[55] to six reactivity ratios $r_{IJ} = k_{II}/k_{IJ}$, I, J = 1, 2, 3 and I ≠ J.

The transition probabilities from the state $- M_I^*$ to the state M_J, I, J = 1, 2 or 3, are computed as:

$$P_{IJ} = \frac{k_{IJ}[-M_I^*][M_J]}{\sum\limits_{J=1}^{3} k_{IJ}[-M_I^*][M_J]} =$$

$$= \frac{k_{IJ}[M_J]}{\sum\limits_{J=1}^{3} k_{IJ}[M_J]}$$

(32)

where [ ] denotes the feed concentration of the monomer. Introducing the reactivity ratios in equation (32) one obtains the nine transition probability expressions as:

$$P_{II} = \left[1 + \frac{[M_J]}{[M_I]} r_{IJ} + \frac{[M_K]}{[M_I]} r_{IK} \right]^{-1}$$

(37)

$$P_{IJ} = \left[1 + \frac{[M_I]}{[M_J]} r_{IJ} + \frac{[M_K]}{[M_I]} \frac{r_{IJ}}{r_{IK}} \right]^{-1}$$

(34)

The Monte Carlo algorithm[38] for simulation of the propagation steps is:

1) specify the reactivity ratios, the monomer feed concentrations, polymerization degree N and the first monomeric unit of the chain, say M_J.

2) compute α, β and γ values as:

$$\alpha_1 = P_{11} \quad , \quad \alpha_2 = P_{11} + P_{12}$$
$$\beta_1 = P_{22} \quad , \quad \beta_2 = P_{22} + P_{21}$$
$$\gamma_1 = P_{33} \quad , \quad \gamma_2 = P_{33} + P_{31}$$

3) generate the random number $\xi \in (0,1)$, ξ's being uniformly repartised within this interval.

4) depending on the concrete value of J (i.e., the last monomeric unit of the chain), test the inequalities:

i) for J = 1:

$\xi < \alpha_1$: monomer M_1 is added;

$\alpha_1 < \xi < \alpha_2$: monomer M_2 is added;

$\alpha_2 < \xi < 1$: monomer M_3 is added;

ii) for J = 2:

$\xi < \beta_1$: monomer M_2 is added;

$\beta_1 < \xi < \beta_2$: monomer M_1 is added;

$\beta_2 < \xi < 1$: monomer M_3 is added;

iii) for J = 3:

$\xi < \gamma_1$: monomer M_3 is added;

$\gamma_1 < \xi < \gamma_2$: monomer M_1 is added;

$\gamma_2 < \xi < 1$: monomer M_2 is added;

Change accordingly the value of J (i.e., the actual last unit of the chain) and resume the step (3).

5) the computing procedure is stopped when the desired polymerization degree is attained.

One may take into account the conversion via (26) - type relations introduced in the transition probability expressions given above.

For methacrylonitrile (M_1)/styrene (M_2)/α - methyl styrene (M_3) terpolymers there are available two sets of reactivity ratios:

(A) r_{12} = 0.44, r_{21} = 0.37, r_{13} = 0.38

r_{31} = 0.53, r_{23} = 1.12, r_{32} = 0.63, and

(B) r_{12} = 0.55, r_{21} = 0.44, r_{13} = 1.49

r_{31} = 0.23, r_{23} = 0.33, r_{32} = 0.57 (according to ref. 56).

The experimental and calculated terpolymers composition is displayed in Table 13.

The Monte Carlo computed sequence distribution for terpolymers no. 4 and 6 from Table 13 is shown in Table 14.

The data collected in Tables 13 and 14 argue for the following two conclusions:

Table 13. Methacrylonitrile (M_1)/styrene (M_2)/α - Methyl Styrene (M_3) terpolymers: composition[*]

No.	Feed composition[a], mole fraction			Terpolymer composition, mole fraction																
				Experimental[a]			Alfrey-Goldfinger equation						Monte Carlo[b]							
							A			B			A			B				
	M_1	M_2	M_3	M_1	M_2	M_3	M_1	M_2	M_3	M_1	M_2	M_3	M_1	M_2	M_3	M_1	M_2	M_3		
1.	0.420	0.307	0.273	0.485	0.284	0.231	0.436	0.303	0.261	0.463	0.315	0.222	0.448	0.304	0.248	0.459	0.335	0.206		
2.	0.555	0.290	0.155	0.544	0.332	0.123	0.525	0.306	0.169	0.557	0.347	0.128	0.533	0.297	0.170	0.562	0.323	0.115		
3.	0.204	0.450	0.286	0.330	0.412	0.259	0.334	0.419	0.247	0.330	0.404	0.266	0.301	0.457	0.242	0.286	0.433	0.281		
4.	0.414	0.424	0.162	0.452	0.389	0.162	0.443	0.403	0.154	0.446	0.402	0.152	0.449	0.405	0.146	0.455	0.407	0.138		
5.	0.180	0.527	0.293	0.265	0.464	0.271	0.264	0.491	0.245	0.250	0.451	0.299	0.273	0.491	0.236	0.255	0.457	0.288		
6.	0.184	0.337	0.480	0.287	0.309	0.405	0.256	0.339	0.405	0.275	0.328	0.397	0.259	0.338	0.403	0.282	0.331	0.387		

a) ref. 56;   b) ref. 38   *) Monte Carlo computations were performed with MEMORY-4 program, available in the chapter 4 of this book.

i) one cannot discriminate among the two sets of reactivity ratios using only the composition data. It is also necessary to consider informations concerning sequence distribution.

ii) the transferability of reactivity ratios from binary copolymerization to ternary one is not assured, but using only composition data it is difficult to achieve a safe conclusion[38].

Table 14. Methacrylonitrile (M_1)/styrene (M_2)/α - Methyl styrene (M_3) terpolymers: sequence distribution (Monte Carlo)

| | Terpolymer no. 4 | | | | | | Terpolymer no. 6 | | | | | |
| | A | | | B | | | A | | | B | | |
m	$n_1(m)$	$n_2(m)$	$n_3(m)$	$n_1(m)$	$n_2(m)$	$n_3(m)$	$n_1(m)$	$n_2(m)$	$n_3(m)$	$n_1(m)$	$n_2(m)$	$n_3(m)$
1	253	222	127	193	246	122	215	183	162	189	227	204
2	60	53	8	74	57	8	19	50	53	34	35	51
3	14	18	1	16	7		2	17	23	7	10	24
4	6	2		10	4		1	14		1	1	1
5	2	3		4	2			2				1
6				1								

The Monte Carlo simulation program developed[37] by Mirabella takes into account the initiation reactions:

$$I\cdot + M_I \xrightarrow{k_{iI}} I - M_I^\cdot \qquad (35)$$

where $I\cdot$ stands for initiation specie. The initiation probabilities are given by:

$$P_{iJ} = \frac{I_J\,[M_J]}{\displaystyle\sum_{J=1}^{3} I_J\,[M_J]} \qquad (36)$$

I_J refers to the relative reactivity of monomer M_J toward the same radical I.

The agreement between copolymer simulation data and experimental data becomes good at polymerization degrees N > 300. Typical results offered by Mirabella's program are displayed in Table 15 (the computer program is available in ref. 37).

Table 15. Styrene (M_1)/butadiene (M_2)/methyl methacrylate (M_3) terpolymers: composition and sequence distribution[a]

N (DP)	Conversion (mole %)	Terpolymer composition (wt%)			Sequence distribution		
		M_1	M_2	M_3	Diads (number fraction of all diads)		Triads (number fraction of all triads)
10	85.0	36.313	22.242	41.444	$M_1M_1 = 0.0703$	$M_2M_2 = 0.1006$	$M_1M_1M_1 = 0.0214$
					$M_1M_2 = 0.1893$	$M_2M_3 = 0.3156$	$M_1M_1M_2 = 0.0259$
					$M_1M_3 = 0.2663$	$M_3M_3 = 0.0579$	$M_1M_1M_3 = 0.0703$
							$M_2M_1M_2 = 0.0392$
							$M_2M_1M_3 = 0.0865$
							$M_3M_1M_3 = 0.0562$
500	85.0	37.539	22.342	40.119	$M_1M_1 = 0.0833$	$M_2M_2 = 0.1051$	$M_1M_1M_1 = 0.0231$
					$M_1M_2 = 0.1749$	$M_2M_3 = 0.3186$	$M_1M_1M_2 = 0.0467$
					$M_1M_3 = 0.2727$	$M_3M_3 = 0.0455$	$M_1M_1M_3 = 0.0737$
							$M_2M_1M_2 = 0.0250$
							$M_2M_1M_3 = 0.0782$
							$M_3M_1M_3 = 0.0604$
1000	85.0	37.588	22.217	40.195	$M_1M_1 = 0.0838$	$M_2M_2 = 0.1045$	$M_1M_1M_1 = 0.0229$
					$M_1M_2 = 0.1734$	$M_2M_3 = 0.3179$	$M_1M_1M_2 = 0.0462$
					$M_1M_3 = 0.2743$	$M_3M_3 = 0.0461$	$M_1M_1M_3 = 0.0755$

Table 15 (continued)

$M_2M_1M_2 = 0.0248$

$M_2M_1M_3 = 0.0776$

$M_3M_1M_3 = 0.0606$

a) The reactivity ratios are: $r_{12} = 0.825$, $r_{13} = 0.485$, $r_{21} = 1.390$, $r_{23} = 0.750$, $r_{31} = 0.422$, $r_{32} = 250$ (ref. 57); the I_j values are: $I_1 = 0.2381$, $I_2 = 0.3095$, $I_3 = 0.4524$ (ref. 58); the above results are averages obtained for 170 chains.

2.4. Reversible copolymerizations.

If it is admitted that the back reaction of equation (37a) is important:

$$- M_I^* + M_J \underset{k'_{IJ}}{\overset{k_{IJ}}{\rightleftarrows}} - M_I M_J^* \tag{37a}$$

then we are dealing with a system which can undergo depolymerization as well as polymerization. Such systems cannot be described simply by reactivity ratios, but one must also introduce the depropagation rate constant, k'_{IJ} as well. This is most easily done by recognizing that the equilibrium constant K_{IJ} for equation (37a) is:

$$K_{IJ} = k_{IJ}/k'_{IJ} \tag{37b}$$

and that the equilibrium constant is related to the temperature , T, and enthalpy, ΔH, and entropy, ΔS, of polymerization by:

$$K_{IJ} = \exp \left(- \frac{\Delta H_{IJ}}{RT} + \frac{\Delta S_{IJ}}{R} \right) \tag{38}$$

Little experimental information exists on the influence of monomer order (i.e., sequence distribution) on the enthalpy or entropy of polymerization, but it is a reasonable assumption that the latter is insensitive to monomer order, since all entropies of polymerization are numerically quite similar. However, enthalpies of polymerization are quite different and may be expected to be strong functions of monomer placement in some cases.

A rigorous kinetic treatment of reversible copolymerization was first given by Lowry[59], making assumptions which were intended for systems such as styrene - α - methylstyrene. The equations derived by Lowry were shown to be valid by a number of workers for several comonomer systems[60-62], but it was recognized that Lowry's equations were quite restricted, in their application, since they described only systems in which some (but not all) of the possible propagation reactions were reversible. A consideration of the probability relationships involved in reversible copolymerization led to an exact solution[63] of the "diad model" (equation 37a) and the "triad model":

$$- M_I M_J{}^* + M_K \rightleftharpoons - M_I M_J M_K{}^* \tag{39}$$

where I, J, K = 1, 2. In principle the probability considerations could generate exact solutions for any length terminal unit, but the increased number of parameters compared to experimental information available does not justify going beyond a triad. The equations derived could, with suitable restrictions, be reduced to those independently derived for any or all models that had been proposed: penultimate effect, Lowry's equations etc.

Monte Carlo simulations of reversible copolymerizations were undertaken by Izu and O'Driscoll[64] to explore the relative importance of the various parameters which control the reaction products. The algorithm used was identical to that given in Section 2.3.1 for irreversible polymerization with a few exceptions:

i) The initial chain consisted of 5 units of the monomer with the higher heat of polymerization.

ii) The possibility of depropagation was explicitly considered in testing the inequality (equation 24) so that the position of ξ in the ranges defined by the two propagation and one depropagation probabilities determined the nature of the next addition or the occurence of depropagation.

iii) The chain was allowed to grow to 500 units or 5000 "events" were allowed to occur, whichever came first. An event was defined as a propagation or depropation reaction.

The results of the Monte Carlo simulation were quite instructive. They showed, for example, that the composition of a chain represented a compromise between kinetic control (as represented by reactivity ratios) and thermodynamic control (as represented by temperature and enthalpies of polymerization). Figure 3 shows the result of one simulation where the effect of temperature is clearly seen: at low temperatures, kinetic control of composition is evident, because the reactivity ratios clearly favour the incorporation of M_2 into the chain. At higher temperature the composition is becoming almost "ideal", refecting the thermodynamic control of the chains' composition.

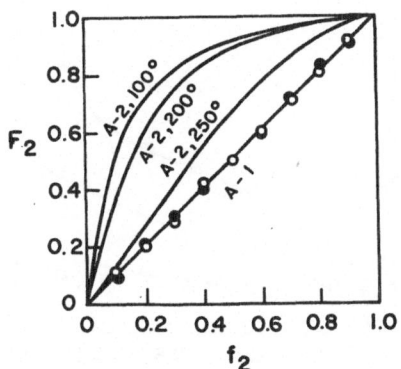

Figure 3. Results of Monte Carlo calculations of copolymer composition
curves: (○) A-1, 100°C; (●) A-2, 200°C.

It is important to recognize that at high temperatures, thermodynamically
unstable sequences can be "frozen in" by the subsequent (kinetically controlled)
additions of monomers which produce a thermodynamically stable end unit.

A number of parametric studies were done which revealed the pattern of composition
behaviours to be expected for different relationships between the enthalpies for
homopolymerization, ΔH_{II}, and those for the cross-propagation steps, ΔH_{IJ}. It is
not possible to deduce any relation from any experimental measurements, since they
can only yield the sum, $(\Delta H_{12} + \Delta H_{21})$, and not their individual values. It was
found that the temperature dependence of the composition behaviour of the simulation for the case where the enthalpy of both cross-propagation steps was the average of the homopolymerization enthalpies:

$$\Delta H_{12} = \Delta H_{21} = \frac{1}{2}(\Delta H_{11} + \Delta H_{22}) \tag{40}$$

was similar to the composition-temperature behaviour which had been experimentally
observed in the copolymerization of methylmethacrylate and α-methylstyrene[65]. The
use of the relation in equation (40) served well to model the composition behaviour
using the analytical expressions previously derived[63]. Figure 4 shows the good
agreement between the experimental data and model predictions. In this instance
the Monte Carlo simulation, through the parametric survey, guided the choice of
the relation to use between the enthalpies in the absence of any other information.

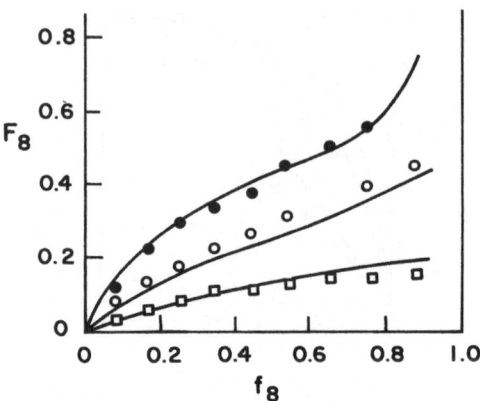

Figure 4. Influence of temperature and feed composition (f) on the copolymer composition (F) in the case of MMA (monomer A) and α-MS (monomer B) copolymerized by free-radical initiation: (●) 60°C; (o) 114°C; (Δ) 147°C; (—) theoretical prediction.

The probability approach used to model reversible copolymerizations has been generalized to n-component polymerizations[66], where n = 2, 3, 4..., and applied to experimental data on the terpolymerization of acrylonitrile, methylmethacrylate and α-methylstyrene. In designing the experiments, the terpolymerizations were simulated by a Monte Carlo program which has been published[32]. The simulations served to define regions of temperature and solvent concentration where significant effects of depropagation could be found when the temperature and/or solvent concentration were changed.

Figure 5 shows the comparision between the Monte Carlo and analytical solutions for the terpolymer compositions expected at various temperatures and monomer feed compositions. Given the complexity of the analytical solutions, the agreement with Monte Carlo simulations was reassuring that the analytical solution were correct.

Figure 6 shows that the analytical solutions were quite capable of properly describing the effects of temperature on terpolymer compositions derived from a given feed. Similar results were observed for the effect of dilution with an inert solvent.

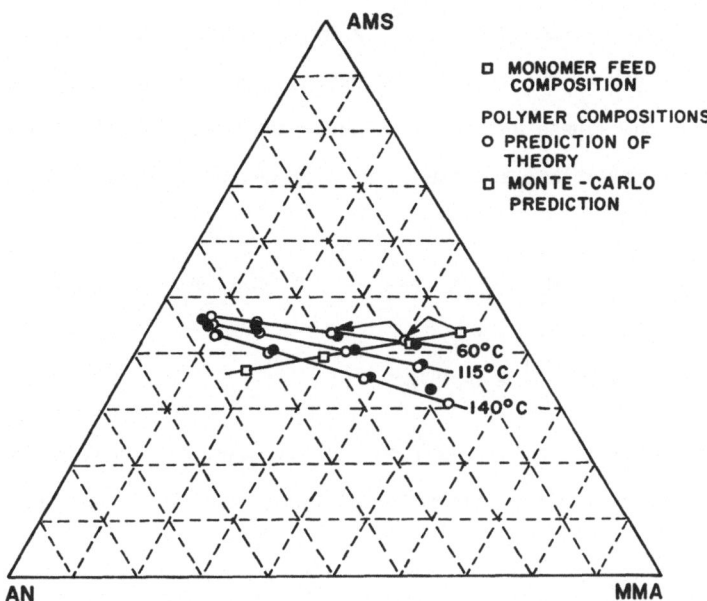

Figure 5. Influence of temperature and feed composition on terpolymer
composition in the case of AN, MMA and AMS terpolymerized in
bulk by free-radical initiation. Arrows are drawn from monomer
feed compositions to polymer compositions.

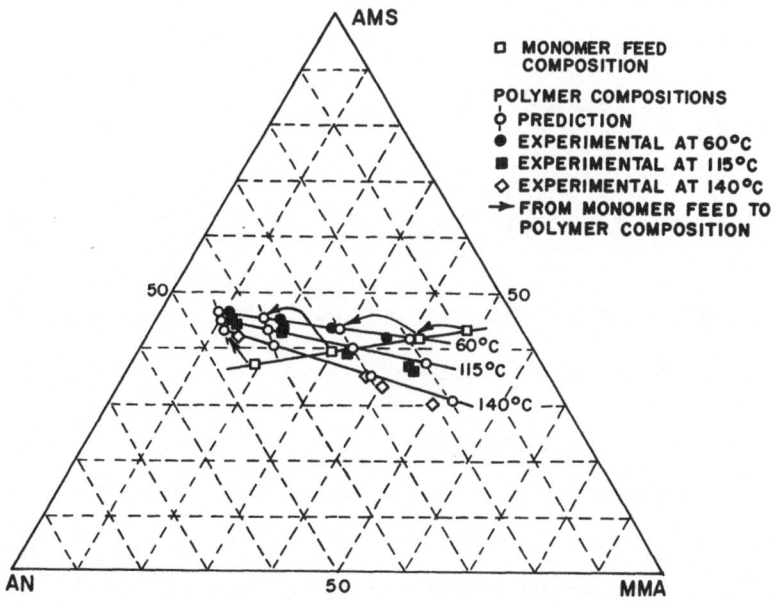

Figure 6. Influence of temperature and feed composition on terpolymer
composition in the case of AN, MMA, and AMS terpolymerized
in bulk by free-radical initiation.

Motoc and Vancea developed[67] a Monte Carlo algorithm which takes into account the length of the terminal sequence, i.e., the sequence which may undergo depolymerization. These authors considered the model in which the propagation steps (41) are reversible and the cross-propagation reactions (42) are irreversible[68]:

$$- M_J(M_I)_{L_I-1} M_I^* + M_I \xrightleftharpoons{k_{II}} - M_J(M_I)_{L_I} M_I^* \quad , I, J = 1, 2; I \neq J \quad (41)$$

$$- M_J M_I^* + M_J \xrightarrow{k_{IJ}} - M_I M_J^* \quad (42)$$

where L_I , I = 1, 2, denotes the length of the terminal sequence.

The model may be extended to the general case without any technical difficulty.

If it is assumed that there exists a minimum length, $L_{I,min}$, for which the reaction (40) occurs, one may write:

$$K_{II} = \begin{cases} 0, \text{ if } L_I < L_{I,min} \\ >0, \text{ if } L_I \geq L_{I,min} \end{cases}$$

The probability that the polymeric chain end $M_J(M_I)_{L_I-1} M_I^*$, where $L_I \geq L_{I,min}$, adds the monomeric unit M_I is:

$$\gamma_I = \frac{k_{II}[-M_I^*][M_I]}{k_{II}[-M_I^*][M_I] + k'_{II} [-M_I^*]} = \frac{[M_I]}{[M_J] + K_{II}^{-1}} \quad (43)$$

and the probability that the newly entered M_I unit undergoes depolymerization is:

$$\overline{\gamma_I} = 1 - \gamma_I \quad , I = 1,2 \quad (44)$$

[ ] stands for feed concentration.

The algorithm is based on that given is Section 2.3.1. for irreversible binary copolymerization with the extensions shown in Figure 7. This algorithm is implemented by the program MEMORY-6 given in chapter 4.

The MEMORY-6 calculation was performed for the reversible copolymerization of α-methyl styrene (M_1) and methyl methacrylate (M_2) at 60°C. The experimental

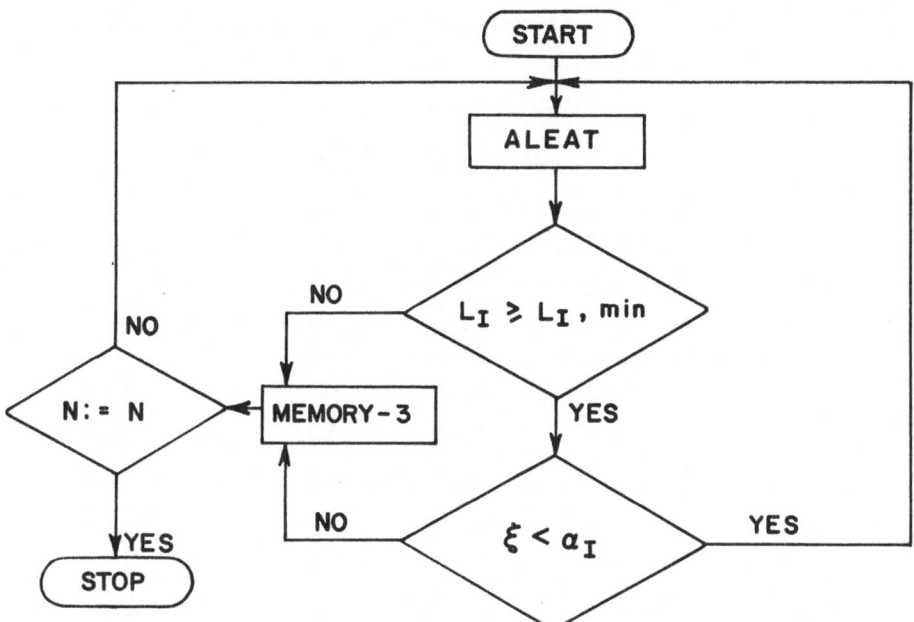

Figure 7. The relation between MEMORY-3 and MEMORY-6 programs.

data were taken from ref. 69 (r_1 = 0.30, r_2 = 0.55, K_1 = k'_{II}/k_{II} = 0.14 mole ℓ^{-1}, K_2 = 0.00) and they were interpreted[69,70] in the sense that in the system α-methyl styrene/methyl methacrylate depolymerization of sequences of two monomer units seems to occur as well as depolymerization of α-methyl styrene from longer sequences.

The Monte Carlo computed copolymer compositions were compared with the figures obtained by applying the corresponding equation developed by Wittmer[70]:

$$\frac{d[M_1]}{d[M_2]} = \frac{1 + r_1 \, ([M_1]/[M_2] - r_1 \, (K_1/[M_2])(1 - x_1))}{1 + r_2 \, ([M_2]/[M_1])} \qquad (45)$$

where $(1 - x_1)$ is given by relation:

$$(1 - x_1) = \frac{r_1 \, ([M_1] + K_1) + [M_2]}{2r_1 K_1} - \left[(\frac{r_1([M_1] + K_1) + [M_2]}{2r_1 K_1})^2 - \frac{[M_1]}{K_1} \right]^{1/2}$$

and Markov chain model proposed by Georgiev[71]. The data are systematized in Table 16.

Table 16.   α-methyl styrene (M_1)/methyl methacrylate copolymers:
composition

M_1 in feed, mole-fraction	M_1 in copolymer, mole - %					
	Experimental	Eq. 45	Markov	Monte Carlo, $L_{I\ min} =$		
				1	2	∞
0.06	9.6	9.58	9.6	9.2	9.2	9.2
0.10	17.7	14.79	14.8	15.3	15.3	15.5
0.20	25.4	25.14	25.1	25.4	25.4	25.7
0.30	33.5	33.06	33.5	32.8	32.8	32.8
0.40	39.7	39.63	39.8	38.5	38.5	38.5
0.50	44.7	45.53	45.1	44.4	44.4	44.7
0.60	52.0	51.31	51.8	51.3	51.3	51.3
0.70	58.7	57.58	58.6	58.0	58.0	58.0
0.80	66.0	65.29	66.5	65.6	65.6	65.9

In this instance, Monte Carlo simulation shows little dependence of
composition on L_I and therefore does not seem to support the existence of reversibility
in the α-methyl styrene/methyl methacrylate system.   The reason is the very small
value assigned to the constant K_1.   Using in simulations a revised value of K_1,
i.e., K_1 = 7.10 mole ℓ^{-1}, (published by the same author[72]), the resultant pattern
of composition revealed it to be quite probable that the depolymerization of
α-methyl styrene sequences takes place _via_ diads as well as longer sequences[73].

2.5.   Configurational sequences in stereoregular polymers.

There is a good, formal analogy between binary copolymerization and
tacticity or configurational placement of monomer units in a homopolymer.   Price
developed[74] the first Monte Carlo algorithm to simulate stereoregular polymers.
The computer program which implements this algorithm is useful in elucidation and
visualization of stereoregular chains.   Both algorithm and program are very
similar with that given in Section 2.3.1. for binary irreversible copolymerization
with a few exceptions:

i. the input is the fractions of isotactic, syndiotactic and heterotactic triplets (denoted by I, S and H, respectively) in the real polymer.

ii. the transition probabilities are:

$P_1 = 2I/(2I + H)$ is the probability that a 0 will follow a 00 to yield a 00.

$P_2 = 2S/(2S + H)$ is the probability that a 0 will follow a 01 to yield a 10.

$P_3 = 1 - P_2$ is the probability that a 0 will follow a 10 to yield a 00.

$P_4 = 1 - P_1$ is the probability that a 0 will follow a 11 to yield a 10.

The symbols 0 and 1 denote the two stereoconfigurations of the chain units.

iii. the program allows the chain to grow to 100 units. Typical results are shown in Figure 8.

000000000000000000001010110101010111111111111111111

111111111111111111111111111111111011111111111111111

a)

100001111111110111000001011000000100000000011110010

011100000010000000011111100010001111111100000000000

b)

Figure 8. a) Results for I = 0.60, H = 0.10, S = 0.30.

b) Results for I = 0.45, H = 0.30, S = 0.25.

This study showed that Monte Carlo simulations are in good agreement with Markov chain model[75,76] with the transition matrix, P, given below:

$$
P = \begin{bmatrix} P_1 & 0 & P_3 & 0 \\ 1-P_1 & 0 & 1-P_3 & 0 \\ 0 & P_2 & 0 & P_4 \\ 0 & 1-P_2 & 0 & 1-P_4 \end{bmatrix}
$$

References

1. F.P. Price, J. Chem. Phys., $\underline{36}$, 209 (1962).

2. E.R. Santee, Jr., V.D. Mochel and M. Morton, J. Polym. Sci, Polym. Lett. Ed., $\underline{11}$, 453 (1973).

3. K. Hatada, Y. Tanaka, Y. Terawaki and H. Okuda, Polym. J., $\underline{5}$, 327 (1973).

4. Y. Tanaka, H. Sato, M. Ogawa, K. Hatada and Y. Terawaki, J. Polym.Sci. Polym. Lett. Ed., $\underline{12}$, 369 (1976).

5. T.K. Wu, D.W. Ovenall and G.S. Reddy, J. Polym. Sci. Polym. Phys. Ed., $\underline{12}$, 901 (1976).

6. J.C. Randall, J. Polym. Sci. Polym. Phys. Ed., $\underline{13}$, 889 (1975).

7. C.J. Carman, Macromolecules, $\underline{6}$, 725 (1973).

8. T.K. Wu and D.W. Ovenall, Macromolecules, $\underline{6}$, 725 (1973).

9. J.C. Randall, "Polymer Sequence Determination", Academic, New York, 1977, chapters 3 and 4.

10. G.E. Ham, ed., "Copolymerization", vol. 1, Interscience, New York, 1964.

11. G.E. Lowry, ed., "Marknov Chains and Monte Carlo Calculations in Polymer Science", Dekker, New York, 1970.

12. B.D. Coleman, J. Polym. Sci., $\underline{31}$, 155 (1958).

13. F.P. Price, J. Chem. Phys., $\underline{36}$, 209 (1962).

14. J. Hijmans, Physica, $\underline{29}$, 1, 819 (1963).

15. A. Rudin, K.F. O'Driscoll and M.S. Rumack, Polymer, in press.

16. K.F. O'Driscoll, J. Polym. Sci. Polym. Chem. Ed., $\underline{18}$, 2747 (1980).

17. H.J. Harwood, J. Polym. Sci., Part C, $\underline{25}$, 37 (1968).

18. H.J. Harwood, Y. Kodaira and D.L. Newman, in "Computers in Polymer Science", eds. J.S. Mattson, H.B. Mark, Jr. and H.C. MacDonald, Jr., Dekker, New York, 1977.

19. I. Motoc, Math. Chem., $\underline{3}$, 245 (1977).

20. R.J. Wilson, "Introduction to Graph Theory", Oliver and Boyd, Edinburgh, 1972.

21. O. Smidstrod and Y. Matsumura, Macromolecules, $\underline{2}$, 42 (1969).

22. P.F. Marconi, R. Tartarelli and M. Capovani, Chemica e l'industria, Suppl. No. 1, $\underline{7}$, 1 (1971).

23. P.F. Marconi, R. Tartarelli and M. Capovani, Chemica e l'industria, Suppl. No. 2, $\underline{7}$, 27 (1971).

24. T. Saito and Y. Matsumura, Polymer J., $\underline{4}$, 124 (1973).

25. I. Motoc, D. Ciubotariu and St. Holban, Rev. Roumaine Chim., $\underline{21}$, 769 (1976).

26. I. Motoc, D. Ciubotariu and St. Holban, Rev. Roumaine Chim., $\underline{21}$, 775 (1976).

27. I. Motoc, D. Ciubotariu and St. Holban, Rev. Roumaine Chim., 21, 1247 (1976).

28. D. Ciubotariu, I. Motoc and St. Holban, Rev. Roumaine Chim., 21, 1253 (1976).

29. I. Motoc and D. Ciubotariu, Rev. Roumaine Chim., 21, 949 (1976).

30. a) I. Motoc, St. Holban and D. Cuibotariu, J. Polym. Sci., Chem. Ed., 15, 1465 (1977); b) 16, 1601 (1978).

31. I. Motoc, R. Vancea and I. Mubcutariu, "Monte Carlo Applications in Chemical Physics of Polymers. II. The Memory Program", Preprint, Solid State Series, 1/1980, Univ. of Timisoara.

32. K.F. O'Driscoll, Computers in Chem. Instrum., 6, 97 (1977).

33. F.M. Mirabella, Polymer, 18, 705 (1977).

34. I. Motoc, R. Vancea and St. Holban, J. Polym. Sci., Chem. Ed., 16, 1587 (1978).

35. I. Motoc, R. Vancea and St. Holban, J. Polym. Sci., Chem. Ed., 16, 1595 (1978).

36. I. Motoc and I. Muscutariu, J. Macromol. Sci., Chem., A15, 75 (1981).

37. F.M. Mirabella, Polymer, 18, 925 (1977).

38. I. Motoc, I. Muscutariu, St. Holban and O. Dragomir, J. Polym. Sci., Chem. Ed., Ed., 18, 1565 (1980).

39. F.R. Mayo and F.M. Lewis, J. Am. Chem. Soc., 66, 1594 (1944).

40. T. Alfrey and G. Goldfinger, J. Chem. Phys., 12, 205 (1944).

41. R.J. Eldridge and F.E. Treloar, J. Polym. Sci., Chem. Ed., 14, 2831 (1976).

42. R. Levins, "Complex Systems", in "Towards a Theoretical Biology", vol. 3, ed. C.H. Waddington, Edinburgh Univ. Press, Edinburgh, 1970.

43. D. Braun, W. Brendlein and G. Mott, Eur. Polym. J., 9, 1007 (1973).

44. J.R. Ebdon, Polymer, 15, 782 (1974).

45. A. Rudin, K.F. O'Driscoll and M.S. Rumack, Polymer, in press.

46. G.E. Ham, J. Polym. Sci., 14, 87 (1954); 54, 1 (1961); 61, 9 (1962).

47. W.E. Barb, J. Polym. Sci., 11, 117 (1953).

48. K. Ito and Y. Yamashita, J. Polym. Sci., A-1, 4, 631 (1966).

49. A. Gilath, S.H. Ronel, M. Shmueli and D.H. Kohn, J. Appl. Polym. Sci., 14, 1491 (1970).

50. E. Merr, T. Alfrey and G. Goldfinger, J. Polym. Sci., 1, 75 (1946).

51. G.W. Snedecor and W.G. Cochran, "Statistical Methods", Iowa State Univ. Press, Ames, Iowa, 1967.

52. M. Kreisel, U. Garbatski and D.H. Kohn, J. Polym. Sci., A2, 105 (1964).

53. E.T. Borrows, R.N. Haward and J. Porges, J. Appl. Chem., 5, 379 (1955).

54. C. Walling and E.R. Brigs, J. Am. Chem. Soc., $\underline{67}$, 1774 (1945).

55. T. Alfrey and G. Goldfinger, J. Chem. Phys., $\underline{12}$, 322 (1944).

56. A. Rudin, S.S.M. Chiang, H.K. Johnson, and P.P. Paulin, Can. J. Chem., $\underline{50}$, 1757 (1972).

57. J. Brandrup and E.H. Immergut, eds., "Polymer Handbook", Wiley, New York, 1975.

58. C. Walling, "Free Radicals in Solution", Wiley, New York, 1957, p. 118-119.

59. G.G. Lowry, J. Polym. Sci., $\underline{42}$, 463 (1960).

60. K.F. O'Driscoll and F.P. Gasparro, J. Macromol. Sci.-Chem., $\underline{A1}$, 643 (1967).

61. K.J. Ivin and R.H. Spensley, J. Macromol. Sci.-Chem., $\underline{A1}$, 653 (1967).

62. Y. Yamashita, H. Kasahara, K. Suyama and M. Okada, Makromol. Chem., $\underline{117}$, 242 (1968).

63. J.A. Howell, M. Izu and K.F. O'Driscoll, J. Polym. Sci. $\underline{A1}$, $\underline{8}$, 699 (1970).

64. M. Izu and K.F. O'Driscoll, J. Polym. Sci., $\underline{A1}$, $\underline{8}$, 1675 (1970).

65. M. Izu and K.F. O'Driscoll, J. Polym. Sci., $\underline{A1}$, $\underline{8}$, 1687 (1970).

66. B.K. Kang, K.F. O'Driscoll and J.A. Howell, J. Polym. Sci., $\underline{A1}$, $\underline{10}$, 2349 (1972).

67. I. Motoc and R. Vancea, J. Polym. Sci., Chem. Ed., $\underline{18}$, 1559 (1980).

68. C.L. Lee, J. Smid and M. Szwarc, J. Am. Chem. Soc., $\underline{85}$, 912 (1963).

69. P. Wittmer, Makromol. Chem., $\underline{103}$, 183 (1967).

70. P. Wittmer, Adv. Chem. Ser., $\underline{99}$, 140 (1971).

71. G.S. Georgiev, J. Macromol. Sci.-Chem., $\underline{10}$, 1063, 1081 (1976).

72. P. Wittmer, Makromol. Chem., $\underline{177}$, 991 (1976).

73. I. Motoc, in preparation.

74. F.P. Price, J. Polym. Sci., C, $\underline{25}$, 3 (1968).

75. F.P. Price, J. Chem. Phys., $\underline{36}$, 209 (1962).

76. F.P. Price, in "Markov Chains and Monte Carlo Calculations in Polymer Science", G.G. Lowry, ed., Dekker, New York, 1970.

Chapter 3

POLYMER CONFIGURATION

W. Bruns

3.1.   Static properties

3.2.   Monte Carlo calculation

      3.2.1.   Outline of the method

      3.2.2.   Chain generation

            3.2.2.1.   Lattice Chains

            3.2.2.2.   Off-lattice chains

      3.2.3.   Test on self-intersection

      3.2.4.   Attrition and enrichment techniques

            3.2.4.1.   Bias Counting

            3.2.4.2.   s-p enrichment technique

            3.2.4.3.   Dimerization method

      3.2.5.   Importance sampling methods

            3.2.5.1.   Bead and subchain rotations

            3.2.5.2.   Slithering snake (reptation method)

      3.2.6.   Multiple chain systems

3.3.   Results

3.1. Static properties

A basic prototype for polymer chains can be represented as

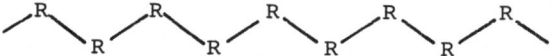

where R is the monomer unit. The configuration shown above is
only one of the enormous number which can be assumed by the chain,
since the molecule can perform internal rotations around each of
the R - R bonds. Because of its flexibility a polymer chain fluctuates
constantly, acquiring various configurations within very short time
invervals. Therefore, at any given moment different chains belonging
to the same ensemble can be found in states having different configurations. Consequently, a polymer described by an ideal chemical
formula is, generally speaking, a mixture of a multiplicity of
different configurations. Thus the problem of the geometry of polymer
chains acquires a statistical character. The chain can be described
therefore only by its averaged parameters. The size and shape of a polymer
chain are of great interest to the polymer scientist, since physical
properties such as light scattering, diffusion, and viscosity can
only be understood if they are discussed in the light of configurational properties.

It is very difficult to describe macromolecular geometries
presisely. During the last decades, however, several quantities
have been defined which may serve to give an impression of the archi-
tecture of the molecule. The (linear) chain is represented by
N + 1 structural units of equal mass numbered from 0 to N. The coordinates of the j-th unit are denoted by $\vec{r}_j$. It is often convenient
to use bond vectors $\{\vec{a}_j\}$, which are defined by

$$\vec{a}_j = \vec{r}_j - \vec{r}_{j-1} = \vec{r}_{j-1,j} \tag{1}$$

One of the above-mentioned quantities is the end-to-end vector $\vec{r} = \vec{r}_{ON}$ which connects the ends of the chain. Ordinarily, only the scalar magnitude of $\vec{r}$ will be of interest:

$$r^2 = (\vec{r}_N - \vec{r}_O)^2 = \sum_{i,j} \vec{a}_i \cdot \vec{a}_j \tag{2}$$

A further quantity is the radius of gyration s. It is defined as the root-mean-square distance from their common centre of gravity of the ensemble of units comprising the chain.

$$s^2 = (N + 1)^{-1} \sum_{i=0}^{N} s_i^2 \tag{3}$$

$\vec{s}_i$ denotes the coordinates of unit i relative to the centre of gravity. The same expression can be obtained for the square of the quasi-radius of gyration s^*, which is defined as the second moment of the segment-density distribution function. This is the probability of finding anyone of the N + 1 units in a given volume element at a given distance relative to the centre of gravity. It is important to note that the higher powers of s and s^* differ .

The value of s_i^2 is usually evaluated by adding the squares of the cartesian coordinates of $\vec{s}_i$, i.e.

$$s_i^2 = s_{xi}^2 + s_{yi}^2 + s_{zi}^2 \tag{4}$$

Accordingly the radius of gyration can also be written as a sum

$$s^2 = s_x^2 + s_y^2 + s_z^2 \tag{5}$$

This decomposition is of course not unique, but depends on the orientation of the coordinate system. If the coordinate axes point in the same direction as the principal axes of inertia of the molecule, the terms of the sum (5) are defined as principal components (l_1^2, l_2^2, l_3^2) of the square radius [2,3]. They provide a good picture

of the shape of the chain. The coincidence of the coordinate axes
and the principal axes of inertia can always be achieved by a proper rotation
of the coordinate system. For this purpose consider the intertia tensor

$$
I = m
\begin{pmatrix}
\sum_i (y_i^2 + z_i^2) & -\sum_i x_i y_i & -\sum_i x_i z_i \\
-\sum_i x_i y_i & \sum_i (x_i^2 + z_i^2) & -\sum_i y_i z_i \\
-\sum_i x_i z_i & -\sum_i y_i z_i & \sum_i (x_i^2 + y_i^2)
\end{pmatrix}
\tag{6}
$$

m is the mass of a chain unit: x_i, y_i, z_i are the cartesian
coordinates of the ith unit relative to the centre of gravity.
Note that $s^2 = \mathrm{tr}\,(I)/\,[2m(N + 1)]$. I (without the factor m) can be
decomposed into the difference of two matrices.

$$
I = (N + 1)\, s^2 E - X
$$

$$
X =
\begin{pmatrix}
\sum_i x_i^2 & \sum_i x_i y_i & \sum_i x_i z_i \\
\sum_i x_i y_i & \sum_i y_i^2 & \sum_i y_i z_i \\
\sum_i x_i z_i & \sum_i y_i z_i & \sum_i z_i^2
\end{pmatrix}
\tag{7}
$$

I is diagonalized by an orthogonal transformation

$$
I_p = Q^{-1} I Q = \mathrm{diag}\,\left[\sum_i (\eta_i^2 + \zeta_i^2),\, \sum_i (\xi_i^2 + \zeta_i^2),\, \sum_i (\xi_i^2 + \eta_i^2)\right]
$$

Since the first expression on the right-hand side is not changed,
and

$$
\sum_i (x_i^2 + y_i^2 + z_i^2) = \sum_i (\xi_i^2 + \eta_i^2 + \zeta_i^2)
\tag{9}
$$

because of the invariance of the trace to similarity transformations,
the second expression becomes

$$Q^{-1}XQ = \text{diag}\ (\ \sum_i \xi_i^2, \sum_i \eta_i^2, \sum_i \zeta_i^2) = (N+1)\text{diag}(l_1^2, l_2^2, l_3^2) \qquad (10)$$

This shows that the eigenvalues of X are identical with the principal components. The column vectors comprising Q give the directions of the principal axes.

Below we give a brief survey of the definitions of some of the quantities described above. Moments of

the end-to-end distance

$$\langle r^{2k}\rangle\ =\ \langle (\vec{r}_N - \vec{r}_O)^{2k}\rangle \qquad (11)$$

the radius of gyration[4]

$$\langle s^{2k}\rangle\ =\ \frac{1}{(N+1)^k}\ \langle (\sum_{i=0}^{N} s_i^2)^k\rangle \qquad (12)$$

the quasi-radius of gyration[4]

$$\langle s^{*2k}\rangle\ =\ \frac{1}{N+1}\ \langle \sum_{i=0}^{N} s_i^{2k}\rangle \qquad (13)$$

By use of the moments it is possible to construct the respective distribution function, or to test roughly a chosen distribution. Several of the higher moments calculated for a freely jointed chain with the length of its bonds fixed and bond angles unconstrained, are given below[5-7]. They may serve for comparison if a new program is to be tested,

$$\langle r^2\rangle_O\ =\ Na^2$$

$$\langle r^4\rangle_O\ =\ \left[\frac{5}{3}N(N-1) + N\right]a^4 \qquad (14)$$

$$\langle r^6\rangle_O\ =\ \left[\frac{35}{9}N(N-1)(N-2) + 7N(N-1) + N\right]a^6$$

$$\langle s^2\rangle_O\ =\ \frac{1}{6}Na^2$$

$$\langle s^4\rangle_O\ =\ \frac{19}{540}N^2 a^4 \qquad\qquad \text{These values are valid}$$

$$\qquad\qquad\qquad\qquad\qquad\qquad\qquad \text{only in the asymptotic} \qquad (15)$$

$$\langle s^6\rangle_O\ =\ \frac{631}{68040}N^3 a^6 \qquad\qquad \text{limit}$$

$$\langle s^{*2}\rangle_O\ =\ \frac{a^2}{6}\ \frac{N}{N+1}\ (N+2)$$

$$\langle s^{*4}\rangle_O\ =\ \frac{a^4}{180}\ \frac{N}{(N+1)^3}\ (10N^4+42N^3+43N^2-3N-2)$$

$$\langle s^{*6}\rangle_O\ =\ \frac{a^6}{68040}\ \frac{N}{(N+1)^5}\ (2030N^7+11452N^6+18089N^5+28900N^4 \qquad (16)$$

$$-2155N^3+5368N^2-8154N+4500)$$

where the subscript O refers to the unperturbed chain.

It is of course also possible to calculate the distribution functions in a more direct way. The procedure is described for the case of the distribution function P(r) of the end-to-end distance $\vec{r}$.

$P(\vec{r})\,\Delta\vec{r}$ ($\Delta\vec{r}$ is a small volume element) is the probability of finding an end-to-end vector $\vec{\rho}$ in the range

$$\vec{r} - \frac{1}{2}\,\Delta\vec{r} \leq \vec{\rho} \leq \vec{r} + \frac{1}{2}\,\Delta\vec{r}$$. P is spherically symmetric, i.e. a function of $r = |\vec{r}|$ only. Hence $P(\vec{r})4\pi\,r^2\,\Delta r$ is the probability of finding an end-to-end distance ρ in the range

$r - \frac{1}{2}\,\Delta r \leq \rho \leq r + \frac{1}{2}\,\Delta r$. In practice, however, one works mainly with the square end-to-end distances. Therefore $P(\vec{r})4\pi\,r^2\,\Delta r$ has to be transformed

$$P(\vec{r})4\,\pi r^2\,\Delta r \;\rightarrow\; P(r^2)\Delta\,r^2 \tag{17}$$

The expression on the right-hand side is the probability of finding the square end-to-end distance ρ^2 in the range

$$r^2 - \frac{1}{2}\,\Delta r^2 \leq \rho^2 \leq r^2 + \frac{1}{2}\,\Delta r^2$$

It can be calculated by taking the ratio of the number of chains, ΔZ_{r^2} , with square end-to-end distances in the range given above, to the total number of chains, Z, comprising the sample. Since $\Delta r^2 = 2r\,\Delta r$, we obtain

$$P(\vec{r}) = \frac{P(r^2)}{2\,\pi r} = \frac{1}{2\,\pi r Z} \quad \frac{\Delta Z_{r^2}}{\Delta\,r^2} \tag{18}$$

By similar considerations the distribution function of a scalar quantity, say s, can be evaluated by

$$P(s) = 2s\,P(s^2) = \frac{2s}{Z} \quad \frac{\Delta Z_{s^2}}{\Delta\,s^2} \tag{19}$$

The configuration of a polymer molecule can also be described by the so-called scattering function $P(\mu)$ (see, for instance, ref. 5, chapter 9). It can be measured directly through elastic radiation scattering experiments. The dependence of the scattered intensity on the scattering angle θ and the wavelength λ is given by

$$P(\mu) = <I(\theta)/I(0)> = (N+1)^{-2} < | \sum_{j=0}^{N} \exp(i\vec{\mu} \cdot \vec{r}_j)|^2 > \qquad (20)$$

with $\mu = (4\pi/\lambda) \sin(\theta/2)$.

The scattering function is very similar to the "structure factor" used in the distribution function theories of liquids. To calculate $P(\mu)$ in a Monte-Carlo experiment, the vector $\vec{\mu}$ is set to $(k,0,0)^T$, $(0,k,0)^T$ or $(0,0,k)^T$ for instance, with various values for k. Since the imaginary parts in Eq. (20) cancel, we obtain

$$P(\mu) = (N+1)^{-2} < (\sum_{j=0}^{N} \cos kl_j)^2 > . \qquad (21)$$

l_j stands for x_j, y_j or z_j and the average has to be taken over all possible configurations. The Monte-Carlo sample has to be generated in such a way that it forms an isotropic system, i.e. the three Eulerian angles must not be preassigned.

3.2. Monte Carlo calculation

3.2.1. Outline of the method

As mentioned before in section 3.1. we are mainly interested in averaged geometrical quantities $< f>$ which can be described in terms of the canonical ensemble

$$<f> = \int f(\{\vec{r}\}) \exp\left[-U(\{\vec{r}\})/(kT)\right] d\{\vec{r}\} / \int \exp\left[-U(\{\vec{r}\})/(kT)\right] d\{\vec{r}\} \qquad (22)$$

f may be r^2, s^2 etc. Before starting the calculation one has to choose a proper model and a potential. There is a broad variety of models which differ in

a) dimensionality (2-, 3-, 4-, 5- ... dimensional models)
b) kind of spatial arrangement

  b1) lattice

   The polymer chain is simulated by a walk in various lattice systems such as cubic, tetrahedral and others.

  b2) off-lattice

   The chains are generated in a continous space. The relative directions of two successive bonds may be chosen either completely at random or with fixed bond angles. In the latter case the angles of rotation (dihedral angles) may be free or restricted.

The total potential energy of a chain is usually split into the potentials corresponding to short-range (constant bond lengths etc.) and long-range interferences. The latter are calculated assuming pairwise additivity of the potentials w between the structural units of the chain

$$W(\{\vec{r}\}) = \sum_{i<j} w(\vec{r}_{ij}) \qquad (23)$$

The following forms of w(r) have mainly been used in the literature

a) $w(r) = 0$ \qquad (random-flight walk, ideal chain)

b) $w(r) = \begin{cases} \infty & r < \sigma \\ 0 & r \geq \sigma \end{cases}$ \qquad hard-sphere-potential

c) $w(r) = \begin{cases} \infty & r < \sigma \\ -\varepsilon & \sigma \leq r < \gamma\sigma \\ 0 & r \geq \gamma\sigma \end{cases}$ \qquad square-well potential

d) $w(r) = 4\varepsilon\left[(\frac{\sigma}{r})^{12} - (\frac{\sigma}{r})^6\right]$ \quad 6-12-Lennard-Jones potential

For lattice chains the potentials a - c correspond to
a) random lattice walks, b) self-avoiding lattice walks (double
occupancy of lattice sites is forbidden), c) similar to b) with
additional consideration of nearest neighbour contacts of nonbonded
chain units.

The method of evaluation of expression (10) depends on the
kind of potential chosen: The importance sampling method described
in section 1.3.2. is always applicable. Since the successive states
of the Markow chain are appreciably correlated, it is difficult to
estimate the standard deviation of the averages. A frequently
used method is the approach of Wood[8]: The total sample consisting
of Z members is broken up into some number l of successive sequences
each with z members. The average value of f is calculated for each
sequence

$$f_s = \frac{1}{z} \sum_{i=(s-1)z+1}^{sz} f(i) \qquad s \in \{1,2,\ldots,l\} \qquad (24)$$

with $Z = lz$

From these we obtain the over-all estimate of $<f>$

$$<f> \simeq \bar{f} = \frac{1}{l} \sum_{s=1}^{l} f_s = \frac{1}{Z} \sum_{i=1}^{Z} f(i) \qquad (25)$$

and its variance

$$\sigma^2 \approx s^2 = \frac{1}{l(l-1)} \sum_{s=1}^{l} (f_s - \bar{f})^2 \qquad (26)$$

s^2, however, depends somewhat on the choice of l.

In the case of random-flight chains and hard-sphere potentials the plain method (see eq. 1.18 for instance) is generally more advantageous. The members of the sample thus generated are independent of each other; this may lead to a faster convergence of the sample averages and permits the estimation of variances by standard statistical methods. Difficulties that may arise in connection with hard-sphere potentials can sometimes be removed by the special techniques which are discussed in section 3.2.4.

3.2.2. Chain generation

Whatever Monte-Carlo method is used, at least one chain configuration has to be generated. If the plain sampling method is used, this generation process has to be repeated very often to obtain a sample of say Z independent configurations. Average statistical-mechanical properties are evaluated by taking the weighted mean of that property with the Boltzmann probability as weighting factor (c.f. Eq. 1.17). If the chains are subject to a hard-sphere potential, the Boltzmann factor can assume only two values: zero, if one or more pairs of structural units have a distance smaller than σ (occupy the same lattice site) or unity, if they do not intersect each other. Here the configuration has to be rejected in the first case, and to be accepted in the latter. In practice, however, one does not generate a complete random chain, followed by tests on intersections, but performs the tests always immediately after the position of a structural unit has been generated. If an intersection occurs the walk is rejected and a new one started. In the next sections the chain generation method is described for lattice and off-lattice chains subject to a hard-sphere potential. Ways of applying importance sampling for chain molecules are treated in Section 3.2.5.

3.2.2.1. Lattice chains

The technique of generating lattice chains has often been described in the literature (see ref. 9 and the references cited there) so that we confine ourselves to a short survey. The origin of the coordinate system is generally taken as the starting point of a walk. The direction of the first step can also be preassigned, say (0,0,1) in a five-choice cubic lattice. Further steps are chosen randomly, either orthogonal to the respective previous steps. or in the same direction. Reverse steps are forbidden, since this would violate the self-avoidance condition. For a walk on a tetra-hedral lattice the first two steps can be arbitrarily chosen, $(0,0,0) \rightarrow (1,1,1) \rightarrow (2,0,2)$. From the last position three choices are allowed. They can be generated by reversing a randomly chosen component of the last step vector. In the example above the last vector is $(1,-1,1)$. The possible next step vectors are $(-1,-1,1)$, $(1,1,1)$, $(1,-1,-1)$, leading to the points $(1,-1,3)$, $(3,1,3)$, $(3,-1,1)$ respectively. The bond length is $\sqrt{3}$ units in this case.

3.2.2.2. Off-lattice chains

The structural units of the chains are considered as spheres of diameter d connected by bonds of length l which are equally distributed in space. The directions of the bonds may be generated at random by an appropriate algorithm (see for example Fig. 1.4). The bond vectors thus obtained are added to construct a chain of the desired bond number[10]. This procedure, however, has two disadvantages: a) It is not possible to generate a chain with fixed bond angles. b) It is not possible to exclude those steps which can be compared with reverse steps in a lattice, i.e. steps which cause an intersection between a structural unit and its second predecessor. The probability of this event has the considerable value of 0.25 in the case of d/l = 1 (touching spheres).

Another method which avoids these shortcomings works as follows[11,12]:

The centre of the zeroth bead is chosen as origin of the coordinate system with the z-axis directed to the first bead. The centre of the second bead is located in the x-z-plane and the bond angle $\pi-\theta_1$ determines its x- and z-coordinates. θ_1 is equally distributed within the limits

$$d^2/2 - 1 \leq \cos \theta_1 \leq 1$$

The lower limit of $\cos \theta_1$ deviates from -1. This ensures that the second bead does not intersect with the zeroth bead. The direction towards the third bead is determined by the supplement of the bond angle θ_2 and the angle of rotation ϕ_2, both being randomly generated. Because the direction of the bonds is intended to be equally distributed over the surface of a sphere, $\cos \theta_2$ and $\phi_2 \in (0,2\pi)$ have to be equally distributed. The range of $\cos \theta_2$ has to be chosen as above for the reasons already discussed. θ_2 and ϕ_2 are angles with regard to an auxiliary coordinate system with its origin in the centre of the first bead. The z-axis coincides with the direction to the second bead and the x-axis is directed in such a manner that angle $\phi_2 = 0$ corresponds to the trans-position of the beads 0,1,2,3. The coordinates of the third bead are (0,0,1) in terms of the auxiliary system attached to the second bead, and must be transformed into the coordinates of the laboratory system by rotation and translation (We have set l = 1).

The proposed position of the third bead is then checked for intersection with the zeroth bead: the chain is discarded in favour of a fresh start from the origin if the distance of the centres of the

two beads happen to be smaller than d. If an intersection does
not occur, further beads are added in the same manner. The coordinates of the (j+1)th bead, $\vec{r}_{j+1}$, may be calculated recursively:

$$\vec{r}_{j+1} = \vec{r}_j + (\prod_{i=1}^{j} A_i) \begin{pmatrix} 0 \\ 0 \\ 1 \end{pmatrix} \quad j = 1,2, \ldots \tag{27}$$

with

$$A_i = \begin{pmatrix} -\cos\theta_i \cos\phi_i & \sin\phi_i & \sin\theta_i \cos\phi_i \\ -\cos\theta_i \sin\phi_i & -\cos\phi_i & \sin\theta_i \sin\phi_i \\ \sin\theta_i & 0 & \cos\theta_i \end{pmatrix}$$

and $\phi_1 = 0$

3.2.3. Test on self-intersection

As mentioned at the beginning of Section 3.2.2., tests on
self-intersection have to be performed if a chain model with a
hard-sphere potential is used. Since the successful addition of a
bead to a chain with k bonds requires k-1 intersection checks,
every chain reaching the desired length of N bonds has undergone a
total of (N-2)(N-3)/2 self-intersection checks during its construction. This number varies as N^2 and is therefore the dominant time-
consuming factor in any Monte-Carlo method. For this reason it
would be a poor procedure to compare the distances of each pair of
beads of a chain with the hard-sphere diameter. In order to cut
down the machine time, various methods have been developed which
are mainly based on hash-code-procedures[10, 11,13-16]. We found
that the computing times were shortest, when we used a
simple testing scheme for short chains (N < 100) and a hash-code
test for longer ones[11]. The simple scheme is based on the fact that
the beads i and j (i $\leq$ j-3) with the coordinates $\vec{r}_i$ and $\vec{r}_j$ intersect

if $|\vec{r}_j - \vec{r}_i| < d$. Necessary (but not sufficient) conditions for the fulfilment of the inequality are $|\Delta x| < d, |\Delta y| < d, |\Delta z| < d$. Hence the complete test above has to be made only if each of these inequalities is satisfied. Several possibilities to set up hash-code tables have been given in the literature (see ref. cited above). Here we describe one of them[11]: Concentric sphere shells of thickness l are put around the origin. If a chain consisting of N+1 beads is to be generated we need N+1 shells, the highest possible number of shells that can be occupied by a stretched chain. Each of the shells is assigned to a place on the list SHELL. "-1" is entered to those "shells" that are empty. A second list INDEX (2N+2) contains the number of the beads (1...N+1) in order, each succeeded by "-1". Before the first entry the two lists have a form shown by Fig. 1a for the example N=9. When shell s is entered by the centre of a bead for the first time, the value of the corresponding place s on list SHELL is changed to the value of the counter (2j-1 if j is the number of the bead) which indicates the place on list INDEX containing the bead number. If shell s is entered for the second (third...) time-say by bead k (value of the counter 2k-1) - place s on SHELL does no longer contain "-1", but the (positive) odd number l. Now we consider the list INDEX. The (l +1)th element may be "-1" or an odd number n. In the first case we change the element to the value of the counter, otherwise we look for the (n+1)th element of INDEX. This may be "-1" or an odd number. In any case we have to go on as before until we find an element "-1", which must then be changed. Fig. 1c shows the state of the lists after the entry of 8 beads, the position of which is shown in Fig. 1b. Let us suppose that this part of the chain is free from intersections. Now a new bead is added. For a test on intersection it must be as- certained in which shell the centre of the bead is positioned. An intersection is only possible with beads belonging to the same shell or the two neighbour shells. These beads can be found by inspection of the two lists. Which beads are in shell s for instance? Place s of SHELL may contain "-1" or an odd number l. In the first case the shell is empty, otherweise position l on list INDEX gives the number of the bead. The quantity on the place l+1 may be "-1" or an odd number m. In the first case no further beads are in shell s, otherwise position m gives the number of the second bead. The quantity on position m+1 indicates whether further beads are in the shell, and we have to go on until we arrive at a "-1". Then all beads in shell s have been found. You may verify this by means of our example.

SHELL		INDEX		bead	shell	SHELL		INDEX	
1	-1	1	1	1	1	1	1	1	1
2	-1	2	-1	2	2	2	3	2	-1
3	-1	3	2	3	2	3	7	3	2
4	-1	4	-1	4	3	4	9	4	5
5	-1	5	3	5	4	5	-1	5	3
6	-1	6	-1	6	3	6	-1	6	13
7	-1	7	4	7	2	7	-1	7	4
8	-1	8	-1	8	3	8	-1	8	11
9	-1	9	5	9		9	-1	9	5
10	-1	10	-1	10		10	-1	10	-1
		11	6	example of a				11	6
		12	-1	configuration				12	15
		13	7					13	7
		14	-1					14	-1
		15	8					15	8
		16	-1					16	-1
		17	9					17	9
		18	-1					18	-1
		19	10					19	10
		20	-1					20	-1

    Fig. 1a   Fig. 1b   Fig. 1c

3.2.4. Attrition and enrichment techniques

When generating a chain having a hard-sphere potential it
often occurs that the non-intersection condition is violated. In this
case the chain has to be discarded in favour of a fresh start. This
has already been discussed in Section 3.2.2. A huge number of
generating attemps have to be made in order to obtain a sufficiently
large sample of non-intersecting chains made up of a large number
of bonds. For a system of freely jointed chains with d/l = 1.0
the half-life is 6.2 steps (bonds). That means at approximately every
6 steps half of the chains must be discarded because of intersections.
As can be shown both empirically and mathematically [17] the decrease
in the number of bonds is very nearly an exponential function of N.
Except for small values of N the attrition resulting from chain
intersections can be charcterized by the expression

$$Z_N = Z_O \exp(-\lambda N) \tag{28}$$

where Z_N is the number of chains remaining after N steps, Z_O is a
constant approximately equal to the number chains originally started,
and λ is the attrition constant depending on the chain model. λ has
the value 0.113 in the case of a freely jointed chain. Other values
can be found for instance in ref. 9. From all that has been said
it becomes clear that enrichment techniques are required to compen-
sate for the attrition. Here we describe only the most frequently
used ones.

3.2.4.1. Bias counting

Rosenbluth and Rosenbluth[18] devised this method which is
mainly suited for lattice models, although it has also been used -
with slight modifications - for off-lattice chains[19]: Assume that

we have a lattice with coordination number q+1. Since a chain is not allowed to double back on itself there are at most q possible subsequent positions at any stage. At each point j in the generation of the chain the q possible directions are examined to determine the directions that lead to a site previously occupied by the mole-cule. Such directions are excluded, and a random choice among the remaining (say q_j) directions is made. This method of generating configurations causes a bias. It can be removed by weighting the statistical-mechanical properties of a configuration with the product

$$\prod_{j=1}^{N+1} (q_j/q).$$

3.2.4.2. s-p Enrichment technique[20]

Z_s chains, each of s bonds, are generated by the plain method. To obtain them approximate number of Z_o starts are necessary. Z_o and Z_s are related by $Z_s = Z_o \exp(-\lambda s)$. Starting with these chains we build up chains with 2 s bonds. In this procedure we would like to overcome the attrition and to have $Z_{2s} \simeq Z_s$. Therefore we use each s-bond chain more then once, say p times. Then Z_{2s} will be appro-ximately given by

$$Z_{2s} = p \ Z_s \ \exp(-\lambda s) = p Z_o \ \exp(-2\lambda s) \tag{29}$$

If this process is continued j times, the number of chains reaching js bonds is

$$Z_{js} = p^{j-1} \ Z_o \ \exp(-\lambda j s) \tag{30}$$

To avoid an explosion in the number of configurations, $p \exp(-\lambda s)$ must be less than unity. A proper choice would be

$$p = [\![v]\!] \qquad v = \exp(\lambda s) \tag{31}$$

The symbol $[\![v]\!]$ means the largest integer that does not exceed the magnitude of v.

We should mention that a theoretical justification of this method cannot be given. Nevertheless this procedure has been used widely and does not seem to yield dubious results. It could be shown, for instance[21], that different choices of s-p pairs do not affect the results of a Monte-Carlo run.

3.2.4.3. Dimerization method [15,22]

This seems to be the most effective procedure to generate very long chains. A large sample of self-avoiding chains (several ten thousand) short enough to make the attrition tolerably small (Z chains with s bonds each) is constructed and committed to a magnetic disc. Next, two of the memorized chains are randomly selected and coupled by fusing the last bead of the first chain with the first bead of the second chain. The orientation of the second half of the "dimer" with respect to the first one is generated by random choices of the three Eulerian angles ($-1 \le \cos\beta \le 1$, $0 \le \alpha, \gamma < 2\pi$). If tests on intersection between the bead of the two partial chains are negative, the data of the chain now consisting of 2s bonds are stored on a second disc after the quantities of interest have been calculated. The dimerization is continued in the same manner up to a sample of Z/2 chains with 2s bonds. In the next stage these in turn are recalled to give Z/4 chains with 4s bonds, etc.

The same criticism applies to this method as to the s-p-enrichment: it cannot be justified theoretically. Alexandrowicz[22] has, however, undertaken several checks the result of which agreed very well with theoretical predictions.

Table 1

d/l	method	sample size	$\langle r^2 \rangle$	$\langle s^2 \rangle$	$\langle r^4 \rangle$	$\langle s^4 \rangle$	$\langle \overset{*}{s}^4 \rangle$
0.1	a	25600	65.25±0.3315	11.00±0.03485	7072±77.44	152.0±1.103	238.4±0.4511
	b	32000	65.19±0.2955	11.05±0.03118	7044±69.56	153.1±0.9887	240.7±0.4102
1.0	a	4000	216.8 ±2.221	33.23±0.1978	66730±1362	1261 ±16.7	1924 ±7.130
	b	32000	216.7 ±0.7949	33.24±0.0698	67170±487.3	1261 ± 5.636	1926 ±2.535

We, in addition, tested the method by a comparision of the data of freely jointed chains with 64 bonds and the ratios $d/l = 0.1$ and 1.0 generated by two ways: a) "in one casting" by the plain method b) by triple dimerization of 8-bond chains. The results (Table 1) show almost perfect agreement, and demonstrate the reliability of the dimerization method.

3.2.5. Importance sampling methods

In this section it is shown how to construct the Markov chain defined by Eq. 1.23. Starting from an arbitrary state of the system a trial state is constructed and tested whether it can be accepted as a new state. This method was initially developed for monatomic systems, where a particle is randomly chosen and shifted within the interval $[x \pm \delta, y \pm \delta, z \pm \delta]$ while all other particles remain in their positions. A reasonable choice for δ is a value leading to rejection of the trial configuration on about 1/2 of the time-steps. Below we describe some methods that have been devised for chain molecules.

3.2.5.1. Bead and subchain rotations

The method of bead rotation was first used for a 5-choice cubic lattice[23], where the angle of two adjacent bonds can be either 90° or 180°. The operation of the program is as follows: One starts with an arbitrary initial configuration. A bead is selected at random along the chain. If this bead is not an end bead, it is moved from its old position $\vec{r}_j$ to a new one $\vec{r}_j{'}$ by

$$\vec{r}_j{'} = \vec{r}_{j-1} + \vec{r}_{j+1} - \vec{r}_j \tag{32}$$

If it is an end bead, its new position is picked at random from the four possibilities satisfying

$$(\vec{r}\,'_{end} - \vec{r}_{next}) \cdot (\vec{r}_{end} - \vec{r}_{next}) = 0 \tag{33}$$

These rules result in no bead movement if bead j-1, j and j+1
lie in a straight line. If the beads lie at three corners of a
square, the jth bead moves across the diagonal of the square to
the opposite corner. An end bead also moves across the diagonal of
a square, in such a way that the bond connecting it with the next-
to-end bead moves through a right angle.

If we work with a freely jointed chain, the movement of the beads
is provided by similar rules as above. If the bead selected at random
is an end bead a new direction of the bond between this bead and
its next neighbour is generated by means of a proper algorithm described in Chapter 1. If this bead is not an end bead, it is moved through
a randomly chosen angle ϕ keeping the bond lengths to its neighbours
fixed. In other words, the vector $\vec{r}_j - \vec{r}_{j-1}$ is rotated by the angle ϕ
about an axis in the direction of the vector $\vec{r}_{j+1} - \vec{r}_{j-1}$. The new
position $\vec{r}_j\,'$ is given by

$$\vec{r}_j\,' = \vec{r}_{j-1} + R(\vec{u},\phi)(\vec{r}_j - \vec{r}_{j-1}) \tag{34}$$

with $\vec{u} = \dfrac{\vec{r}_{j+1}-\vec{r}_{j-1}}{|\vec{r}_{j+1}-\vec{r}_{j-1}|} = (c_1, c_2, c_3)^T$

and $R(\vec{u},\phi) = \begin{pmatrix} c_1^2+(1-c_1^2)\cos\phi & c_1c_2(1-\cos\phi)-c_3\sin\phi & c_1c_3(1-\cos\phi)+c_2\sin\phi \\ c_1c_2(1-\cos\phi)+c_3\sin\phi & c_2^2+(1-c_2^2)\cos\phi & c_2c_3(1-\cos\phi)-c_1\sin\phi \\ c_1c_3(1-\cos\phi)-c_2\sin\phi & c_2c_3(1-\cos\phi)+c_1\sin\phi & c_3^2+(1-c_3^2)\cos\phi \end{pmatrix}$

This method has been used by Baumgärtner and Binder[24]. It has, however,
to be modified in the case of chains with fixed bond angles[25].

Here a bead j is selected at random and all beads numbered j+1 through N inclusive are rotated through circular arcs of the random angle ϕ about the extension of the bond connecting the beads j-1 and j. By this means a new point in conformation space is reached corresponding to a polymer which differs from the previous polymer in exactly one dihedral angle, but which can be considerably different from it in overall shape and properties. Eq. 34 can be used similarly to calculate the new positions.

3.2.5.2. Slithering snake (reptation method)[26, 27]

This method works similarly for lattice and off-lattice chain models. Starting with an arbitrary configuration, one end of the chain is selected at random. The generation of the trial configuration consists of adding a new bead with a randomly chosen direction, and allowing the rest of the chain to move forward along the old contour. The site that has previously been occupied by the "tail" of the chain is left vacant. The change is allowed to take place, and the new configuration is accepted if the relevant tests are met. The choice of the random direction is obvious in the case of lattice models and freely jointed chains. If the model has fixed bond angles, the direction of the new bond can be obtained by rotation of the vector $\vec{r}_{n-1} - \vec{r}_{n-2}$ about the vector $\vec{r}_n - \vec{r}_{n-1}$ by randomly chosen angle ϕ.

3.2.6. Multiple chain systems

So far we have only considered the methods of generating single chain molecules. A multiple chain simulation, however, is necessary if thermodynamic properties (the equation of state for instance) or the effect of density on the average chain configuration are studied. The simulation starts with the construction of a set of Z chains with N bonds each. The three Eulerian angles are

chosen randomly and the rotated chains introduced into a hypothetical
cubic box (side length L) by a proper random translation. In order
to simulate more closely the behaviour of an infinite system and
to minimize surface effects it is customary to use so-called
"periodic boundary conditions":The box containing the Z chains is
surrounded on all sides by periodically repeating replicas of itself
with the chains in each image box in the same relative position. If
a chain element leaves the box on one side, another chain element
enters on the opposite side thus holding the number of chain elements
in each box constant. The calculation of the energy of interaction of
a particle with its neighbours is made with the help of the minimum
image convention[29]: A particle i lying within the basic box is
assumed to interact only with the nearest image of any other particle j
in the box. The interaction is set to zero if this distance is greater
then $r_c \leq L/2$ (cutoff radius). The significance of this restriction is
that i can interact with one image of j at most.
A simple algorithm achives the intended effect:
With the origin of the coordinate system in the centre of the basic
box we obtain for $\vec{r}_{ij} = (x_{ij}, y_{ij}, z_{ij})^T$

$$v \leftarrow [2x_{ij}/L], \quad x_{ij} \leftarrow x_{ij} - vL \tag{33}$$

The symbol $[\ldots]$ has the same meaning as in Eq. 29 , similar equations hold for y_{ij} and z_{ij}.
The minimum image convention cannot be used of course if the potential is very long ranged, as in the case of polyelectrolytes. Instead, the periodicity of the system can be exploited in order to
calculate energies and forces by methods widely used in solid state
physics for the calculation of lattice sums[30]. But we shall not concern
ourselves with this subject.

3.3. <u>Results</u>

The primary objective of the first Monte Carlo investigation
on polymers[31] was to characterize the behaviour of the characteristic
ratio, $<r^2>/N$, for large value of N, if the chains are subject to
the excluded volume effect. At that time both convergence[32-35]
and divergence[36,37] of this ratio had been predicted for three-dimensional chains. The first publications in this field[38] showed ambi-
guous results because only small samples could be generated. Other
investigations, however, which were published subsequently, showed
that the dependence of the mean-square end-to-end distance on the
number of bonds N can be described by the following simple law
expression

$$<r^2> = aN^{\gamma} \tag{34}$$

where a and γ are constants.
This equation can be reconciled with intrinsic viscosity formulas
of the type $[\eta] = KM^a$. It holds for any lattice or non-lattice
model of any dimensionality, provided N is sufficiently large.
A similar equation is valid for $<s^2>$. Domb studied the behaviour
of self-avoiding walks on various 2-and 3-dimensional lattice
types by an exact enumeration method[39]. On the basis of his results,
which agreed satisfactorily with Monte-Carlo data of other authors,
he suggested that γ should be a universal constant depending only
on the dimensionality of this model:

$$\gamma = \begin{cases} 3/2 & \text{2 dimensions} \\ 6/5 & \text{3 dimensions} \end{cases}$$

This implied a confirmation of the theoretical work of Flory[36],
Edwards[40], and Reiss[41], who predicted the asymptotic form

$$\lim_{z \to \infty} \alpha_r^5 \sim z \tag{35}$$

Since $\alpha_r^5 \sim <r^2>$ and $z \sim N^{1/2}$ we reobtain Eq.(34) with $\gamma = 6/5$.
At the same time, however, some Monte Carlo studies have been
published which do not obtain the parameter γ as a universal constant.
They involve finite intermolecular interactions between nonbonded
nearest neighbours[42-44], branched systems[45], varying excluded
volume[46], and off-lattice studies[11,14,47-50]. Some results are
given in Table 2 and a graphical representation in Fig. 2.

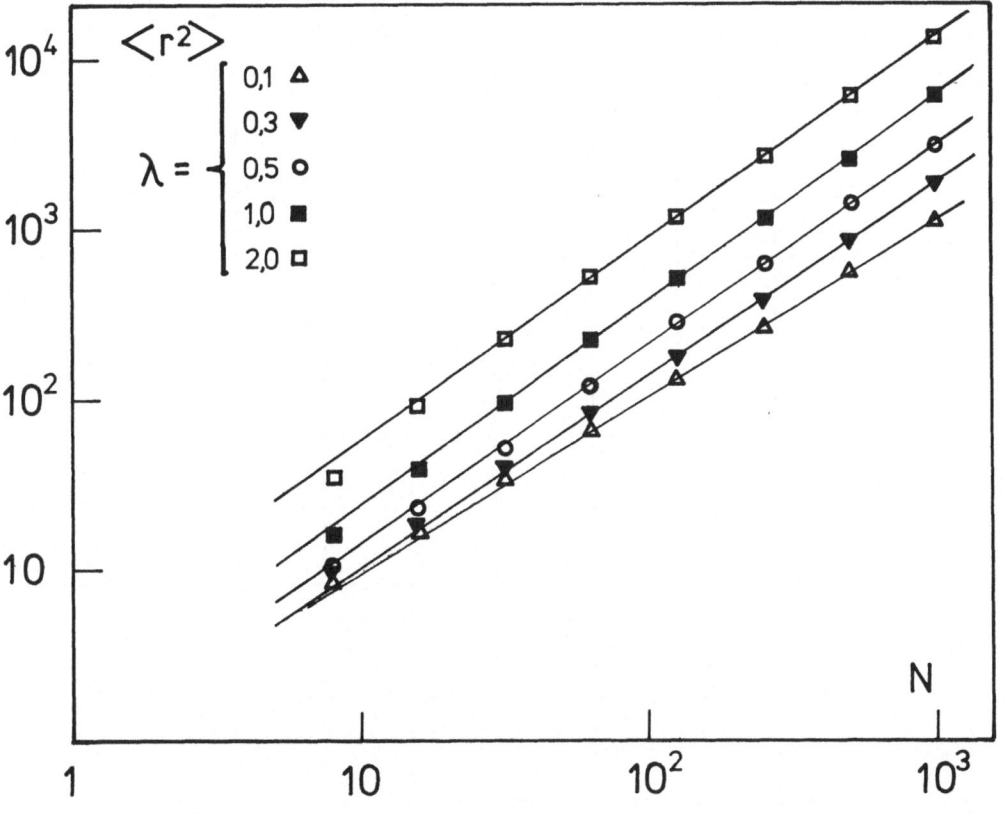

Fig. 2

Table 2

N \ Y	0.1	0.3	0.5	1.0	2.0
8	8.071 ± 0.01215	8.594 ± 0.01251	10.23 ± 0.01348	15.85 ± 0.01619	34.64 ± 0.02137
16	16.25 ± 0.03587	17.91 ± 0.03777	23.02 ± 0.0439	38.58 ± 0.06234	91.89 ± 0.1184
32	32.51 ± 0.1038	37.76 ± 0.1138	52.19 ± 0.1428	92.49 ± 0.2276	224.6 ± 0.5051
64	65.19 ± 0.2955	80.52 ± 0.3413	118.7 ± 0.4634	216.7 ± 0.7949	524.7 ± 1.875
128	130.5 ± 0.8384	174.9 ± 1.056	272.9 ± 1.538	502.3 ± 2.690	1194 ± 6.448
256	266.6 ± 2.444	382.9 ± 3.290	620.4 ± 4.925	1156 ± 8.891	2696 ± 20.82
512	533.4 ± 6.860	853.1 ± 10.20	1416 ± 15.83	2648 ± 29.38	6016 ± 65.74
1024	1077 ± 19.71	1890 ± 31.30	3169 ± 50.85	6168 ± 97.88	13500 ± 209.2

As can easily be verified by our data, γ is not constant but depends on the ratio λ of bead diameter to bond length. This seems to contradict the idea of universality which emerged during the investigation of critical point transitions.

Wilson[51] has introduced the renormalization group approach which has provided a general theoretical frame work for understanding the characteristic features of critical point behaviour, and de Gennes[52] has pointed out that a special case of the renormalization group corresponds exactly to the problem of a self-avoiding walk on a lattice. But the Monte Carlo results described above are incompatible with the concept of universality. A possible explanation could be: The chains generated by a Monte-Carlo run are too short to reach their asymptotic behaviour. Flory and Fisk[53], for instance, estimated the range, required for convergence of $<r^2>/N^{6/5}$ to be of order of 10^6 steps. This requirement, however, cannot be satisfied by numerical methods at present.

Recently new attempts[54,55] have been made to calculate the value γ by a Monte-Carlo renormalization group procedure. The authors presumed the validity of the scaling and universality concepts, and developed a renormalization group analysis, which is analogous to the real-space renormalization group transformation of spin systems. In this technique only relatively short chains are necessary to calculate γ. The authors estimated it as $\gamma = 1.18 \pm 0.02$ in the asymptotic limit for any hard-sphere potential.

The mean-square end-to-end distance depends on the number of bonds N and an excluded-volume parameter β. According to the theories mentioned above these two parameters always appear in the form $\beta N^{1/2}$. The validity of this statement has been checked by several authors by means of lattice models[15,56,57]. The results have been found to be consistent with the theoretical predictions. Attempts have been made to confirm this on the basis of an off-lattice model[11,49,50], but the agreement was only poor. The reason for these contradictory outcomes lies in the different potentials employed. For the calculations on the lattice model an interaction parameter $w = 1 - \beta/a^3$ ($0 \leq w \leq 1$) has been associated with each self-intersection of the chain. This is equivalent to the application of a pseudopotential with the three-dimensional Dirac delta function which is usually considered by theory. It is, however, impossible to introduce such a potential into an off-lattice model, and it has been shown[58] that β cannot be defined uniquely if potentials other than the pseudopotential combined with finite values of N are used. The introduction of a constant value for β (for example the eightfold of a bead volume)must, therefore, give results that can only yield a qualified comparision with the findings of theory. The disposition of segments of a chain molecule about its centre of mass must on the long-time average be spherically symmetrical in space, but this will in general not be true of an instantaneous chain information. It is possible to get some information about the shape of the chain if the principal components of the radius of gyration $l_1 \leq l_2 \leq l_3$ are evaluated (see section 3.1). The results published for lattice[59-62] and off-lattice[3] models are in good agreement:

$$\langle l_1^2 \rangle : \langle l_2^2 \rangle : \langle l_3^2 \rangle = 1 : 2.7 : 11.9 \text{ (random flight chains)}$$

and

$$\langle l_1^2 \rangle : \langle l_2^2 \rangle : \langle l_3^2 \rangle = 1 : 3.0 : 14.6 \text{ (self-avoiding chains/} \lambda = 1.0)$$

These values indicate strong departures from spherical symmetry for the random flight chain which become even more pronounced if the excluded-volume effect is taken into account.

The distribution function of the end-to-end distance has been proposed[63] to be of the form

$$P(r)dr = Cr^{\epsilon} \exp\left[-(r/\sigma)^t\right]dr. \qquad (36)$$

C and σ are determined by the conditions

$$\int_o^{\infty} W(r)dr = 1 \quad \text{and} \quad \int_o^{\infty} r^2 W(r)dr = \langle r^2 \rangle$$

Several attempts[4,64-66] have been undertaken to verify Eq. 36 and to evaluate the parameters ε and t. The distribution seems to be best described with $\epsilon = t = 5/2$. It should be mentioned in this connection that distribution functions other than Eq. 36 have been proposed and investigated [65,67-69], but we do not discuss them further.

Most of the Monte-Carlo calculations that have been published so far have been carried out on single nonintersecting random walks in various lattice systems. Since the 1960s this model has been improved in many respects, depending on the problems that are be treated.

a) Other than hard-sphere potentials[70-76] (mainly of square-well type) have been introduced so that the effect of temperature on the size and shape of the molecule as well as thermodynamic properties could be investigated. Among them an important quantity is the so called θ-point[5]. It has two definitions which are not necessarily consistent:

(1) the point at which an isolated polymer molecule behaves as a random coil and (2) the point at which the second virial coefficient A_2 of the polymer solution vanishes. The long chain limit has to be taken for both definitions. Some values of the reduced θ-point that have been published agree surprisingly

well despite of the diversity of the models, potentials and definitions used for their evaluation: (1)[74] simple cubic lattice, square-well potential, $<r^2> \sim N$; (2)[76], freely jointed chain, Lennard-Jones potential, $\beta = 0$;

(3)[77] simple cubic lattice, square-well potential, $A_2 = 0$. In each case $\varepsilon/(k\theta) = 0.27$.

The θ-point can be viewed as the state of the system at which the repulsive energies are balanced by attractive energies so that the polymer behaves as if it were a random chain. The θ-temperature corresponds, therefore, to the Boyle-temperature of real gases.

If the attractive energies are incrased one might expect the chain to finally collapse upon itself. Such a transition is suggested by inspection of the $<s^2>/N$ vs. $\varepsilon/(kT)$ plot of McCrackin et al[74]. Their curves showed a steepness near the theta region, which increased with increasing N, so that a discontinuity may be expected for long enough chains.

b) If the properties of a distinct class of polymer molecules are investigated the models described so far are too approximate. Some authors[78-87], therefore, have used more complex and realistic polymer models with emphasis on polypeptide models in their Monte-Carlo studies. Their results are, however, too specific to the respective model to be discussed here. The interested reader may consult the references cited above.

c) Intra- as well as intermolcular interactions are involved if a bulk polymer or a concentrated solution is considered. These cannot of course be simulated by a single model chain. Since the change of the mean dimension of a polymer chain with concentration is a challenging problem, a lot of studies[88-94] have been made on this subject by means of computer simulations

of multiple chains systems. It has always been observed that
the mean dimensions decrease steadily as the concentration
increases. The limiting values of the shape parameters for bulk
polymers appear to correspond to the values for random walks
without step reversals. No intra- or intermolecular orientation
differing from the random one has been found[95-97] in the
amorphous bulk state.

d)   Other properties that have been studied are the behaviour of
a chain subject to special constraints[98-100] or absorbed at a
surface[101, 102].

A further interesting application of the Monte-Carlo method is
the study of chain dynamics[103-113]. The so-called molecular dynamics
method based on the solution of the classical equations of motion
may be thought to be the better tool for that purpose, but this
method suffers from long computation times. The time, which is as
such not a variable in Monte-Carlo calculations can easily be introduced if an importance sampling method via a Markov process is
employed (see section 3.2.5.).

The process: (1) the random selection of a bead, (2) the trial
replacement of this bead to a new positon, and (3) the test whether
the old or the new configuration has to be accepted, is usually
called a "bead cycle". It is reasonable and in accordance with the
definition in molecular dynamics calculations to consider N "bead
cycles" as the time unit if N is the total number of beads comprising the system. For a comparison with real systems this time
unit can be treated as an adjustable parameter. The chain relaxation behaviour has been studied with this method and expressed in
terms of time-correlation functions. These play the dominant rôle in
the theory of linear irreversible processes, just as partition
functions play the dominant rôle in equilibrium statistical

mechanics. The autocorrelation functions may be defined as

$$C(\vec{v},t) = \frac{\langle v(t_o) \cdot v(t_o + t)\rangle}{\langle v^2\rangle_e} \qquad \text{for vector quantities,}$$

and

$$C(f,t) = \frac{\langle f(t_o)f(t_o + t)\rangle - \langle f\rangle_e^2}{\langle f^2\rangle_e - \langle f\rangle_e^2} \qquad \text{for scalar ones.}$$

$\langle \ldots \rangle_e$ are equilibrium quantities.

References

1. H. Yamakawa, Modern Theory of Polymer Solutions, Chap. 1 (Harper & Row, New York, 1971).

2. K. Šolc, J. Chem. Phys. 55, 335 (1971)

3. W. Bruns, Colloid & Polymer Sci. 254, 325 (1976)

4. W. Bruns, Makromol. Chem. 158, 177 (1972)

5. P.J. Flory, Statistical Mechanics of Chain Molecules, Chap. 8 (Interscience Publ., New York 1969)

6. M. Fixman, J. Chem. Phys. 36, 306 (1962)

7. W. Bruns, Makromol. Chem. 170, 191 (1973)

8. W.W. Wood, Monte Carlo Calculation of the Equation of State of Systems of 12 and 48 Hard Circles. Los Alamos Scientific Laboratory Report LA-2827 (Los Alamos, New Mexico, Juli 1, 1963)

9. S. Windwer in "Markov chains and Monte Carlo Calculations in Polymer Science, ed. by G.G. Lowry, Marcel Dekker, New York 1970

10. R. Grishman, J. Chem. Phys. 58, 220 (1973)

11. W. Bruns, J. Phys. A: Math. Nucl. Gen. 10, 1963 (1977); Makromol. Chem. 124, 91 (1969); ibid. 134, 193 (1970); ibid. 176, 813 (1975)

12. W. Bruns and J. Naghizadeh, J. Chem. Phys. 65, 747 (1976)

13. P.J. Gans, J. Chem. Phys. 42, 4159 (1965)

14. R.J. Fleming, Proc. Roy. Soc. 90, 1003 (1967)

15. Z. Alexandrowicz and Y. Accad, J. Chem. Phys. 54, 5338 (1971)

16. P.C. Jurs and J.E. Reissner, J. Chem. Phys. 55, 4948 (1971)

17. J.M. Hammersley and K.W. Morton, J. Roy. Statistical Soc. B 16, 23 (1954)

18. M.N. Rosenbluth and A.R. Rosenbluth, J. Chem. Phys. 23, 256 (1955)

19. N.C. Smith and R.J. Fleming, J. Phys. A: Math. Gen. 8, 929 (1975)

20. F.T. Wall and J.J. Erpenbeck, J. Chem. Phys. 30, 634 (1959)

21. W. Bruns and L. Vogel, unpublished results

22. Z. Alexandrowicz, J. Chem. Phys. 51, 561 (1969)

23. P.H. Verdier and W.H. Stockmeyer, J. Chem. Phys. 36, 227 (1962)

24.   A. Baumgärtner and K. Binder, J. Chem. Phys. 71, 2541 (1979)

25.   S.D. Stellman and P.J. Gans, Makromolecules 5, 516 (1972)

26.   A.K. Kron, Vysokomol. soyed. 7, 1228 (1965)

27.   F.T. Wall and F. Mandel, J. Chem. Phys. 63, 4592 (1975)

28.   J.G. Curro, J. Chem. Phys. 61, 1203 (1974)

29.   W.W. Wood and F.R. Parker, J. Chem. Phys. 27, 720 (1957)

30.   S.G. Brush, H.L. Sahlin, and G. Teller, J. Chem. Phys. 45, 2102 (1966)

31.   F.T. Wall, L.A. Hiller, Jr., and D.J. Wheeler, J. Chem. Phys. 22, 1036 (1953)

32.   J.J. Hermans, Rec. Trav. chim. 69, 220 (1950)

33.   J.J. Hermans, M.S. Klamkin, and R. Ullman, J. Chem. Phys. 20, 1360 (1952)

34.   R. Ullman and J.J. Hermans, J. Polymer Sci. 10, 559 (1953)

35.   F.T. Wall, J. Chem. Phys. 21, 1914 (1953)

36.   P.J. Flory, J. Chem. Phys. 17, 303 (1949)

37.   H.M. James, J. Chem. Phys. 21, 1628 (1953)

38.   F.T. Wall, L.A. Hiller, Jr., and W.F. Atchinson
       J. Chem. Phys. 23, 913 (1955); ibid. 23, 2314 (1955);
       ibid. 26, 1742 (1957)

39.   C. Domb, J. Chem. Phys. 38, 2957 (1963)

40.   S.F. Edwards, Proc. Phys. Soc. (London) 85, 613 (1965)

41.   H. Reiss, J. Chem. Phys. 47, 186 (1967)

42.   F.T. Wall and J. Mazur, see ref. 70

43.   F.T. Wall, S. Windwer, and P.J. Gans, J. Chem. Phys. 38, 2220 (1963)

44.   P. Mark, S. Windwer, J. Chem. Phys. 47, 708 (1967)

45.   L.V. Gallacher and S. Windwer, J. Chem. Phys. 44, 1139 (1966)

46.   A. Rice and S. Windwer, J. Chem. Phys. 43, 3773 (1965)

47.   S. Windwer, J. Chem. Phys. 43, 115 (1965)

48.   E. Loftus and P.J. Gans, J. Chem. Phys. 49, 3828 (1968)

49.   N.C. Smith and R.J. Fleming, J. Phys. A: Math. Gen. 8, 729 (1975)

139

50. R.J. Fleming, J. Phys. A: Math. Gen. $\underline{12}$, 2157 (1979)

51. K.G. Wilson and J.B. Kogur, Phys. Rep. (C) $\underline{12}$, 75 (1974)

52. P. de Gennes, Phys. Lett. (A) $\underline{38}$, 339 (1972)

53. P.J. Flory and S. Fisk, J. Chem. Phys. $\underline{44}$, 2243 (1966)

54. A. Baumgärtner, J. Phys. A: Math. Gen. $\underline{13}$, L39 (1980)

55. K. Kremer, A. Baumgärtner, and K. Binder (in press)

56. A.K. Kron and O.B.Ptitsyn, Vysokomol. soyed. $\underline{6}$, 862 (1964)

57. M. Lax and J. Gillis, Macromolecules $\underline{10}$, 334 (1977)

58. W. Bruns, J. Chem. Phys. $\underline{73}$, 1970 (1980)

59. K. Šolc and W.H. Stockmayer, J. Chem. Phys. $\underline{54}$, 2756 (1971)

60. D.E. Kranbuehl, P.H. Verdier, and J.M. Spencer, J. Chem. Phys. $\underline{59}$, 3861 (1973)

61. J. Mazur, C.M. Guttman, and F.L. McCrackin, Macromolecules $\underline{6}$, 872 (1973)

62. R.J. Rubin and J. Mazur, Macromolecues $\underline{10}$, 139 (1977)

63. C. Domb, J. Gillis, and G. Willmers Proc. Phys. Soc. London $\underline{85}$, 625 (1965)

64. F.T. Wall and F.T. Hioe, J. Phys. Chem. $\underline{74}$, 4410 (1970)

65. S.D. Stellman and P.J. Gans, Macromolecues $\underline{5}$, 720 (1972)

66. R. Grishman, J. Chem. Phys. $\underline{58}$, 5309 (1973)

67. T.F. Schatzki, J. Polymer Sci. $\underline{57}$, 337 (1962)

68. F.T. Wall and R.A. White, Macromolecules $\underline{7}$, 849 (1974)

69. F.T. Wall and W.A. Seitz, J. Chem. Phys. $\underline{67}$, 258 (1977)

70. F.T. Wall and J. Mazur, Ann. N.Y. Acad. Sci. $\underline{89}$, 608 (1961)

71. F.T. Wall, S. Windwer, and P.J. Gans, J. Chem. Phys. $\underline{38}$, 2228 (1963)

72. J. Mazur and F.L. McCrackin, J. Chem. Soc. $\underline{49}$, 648 (1968)

73. R.G. Kirste, Disc. Faraday Soc. $\underline{49}$, 51 (1970)

74. F.L. McCrackin, J. Mazur, and C.M.Guttman, Macromolecules $\underline{6}$, 859 (1973)

75. A.T. Clark and M. Lal, Brit. Polym. J. $\underline{9}$, 92 (1977)

76. A. Baumgärtner, J. Chem. Phys. $\underline{72}$, 871 (1980)

77.   M. Janssens and A. Bellemans, Macromolecules 9, 303 (1976)

78.   K. Suzuki and Y. Nakata Bull. Chem. Soc. Japan 43, 1006 (1970)

79.   K.K. Knaell and R.A. Scott III, J. Chem. Phys. 54, 566, 3556
       (1971)

80.   H.E. Warvari, K.K. Knaell, and R.A. Scott III, J. Chem. Phys.
       55, 2020 (1971); ibid. 56, 2903 (1972); ibid. 57, 1161 (1972)

81.   H.E. Warvari and R.A. Scott III, J. Chem. Phys. 57, 1146,
       1154 (1972)

82.   S. Tanaka and A. Nakajima, Macromolecules 5, 708, 714 (1972)

83.   S. Preliminat and J. Hermans, Jr., J. Chem. Phys. 59, 2602
       (1973)

84.   Y. Nakata and K. Suzuki, Polymer J. 6, 242 (1974)

85.   D.E. Neves and R.A. Scott III, Macromolecules 8, 267 (1975);
       ibid. 9, 554 (1976); ibid. 10, 339 (1977)

86.   M. Sisido, Y. Imanishi, and T. Higashimura, Macromolecules
       9, 320, 389 (1976)

87.   N. Go and H.A. Scheraga, Macromolecules 11, 552 (1978)

88.   S. Bluestone and M.J. Vold, J. Polymer Sci. A2, 289 (1964);
       J. Chem. Phys. 42, 4175 (1965)

89.   J.G. Curro, J. Chem. Phys. 61, 1203 (1974); ibid 64, 2496
       (1976); Macromolecules 12, 467 (1979)

90.   E. de Vos and A. Bellemans, Macromolecules 7, 812 (1974);
       ibid. 8, 651 (1975)

91.   T.M. Birshtein, A.M. Skvortsov, and A.A. Sariban, Vysokomol.
       soyed. A 19, 63 (1977)

92.   F.T. Wall, J.C. Chin, and F. Mandel, J. Chem. Phys. 66,
       3143 (1977)

93.   F.T. Wall and W.A. Seitz, J. Chem. Phys. 67, 3722 (1977)

94.   H. Okamoto, J. Chem. Phys. 70, 1690 (1979)

95.   H. Okamoto, J. Chem. Phys. 64, 2686 (1976)

96.   A.M. Skvortov, A.A. Sariban, and T.M. Birshtein, Vysokomol.
       soyed. A 19, 1014 (1977)

97.   R. de Santis and H.G. Zachmann, Colloid a. Polymer Sci.
       255, 729 (1977).

98.   F.T. Wall, F. Mandel, and J.C. Chin, J. Chem. Phys. 65,
      2231 (1976)

99.   F.T. Wall, J.C. Chin and F. Mandel, J. Chem. Phys. 66, 3066
      (1977)

100. J.B. Smitham and D.H. Napper, J. Polymer Sci., Symposium
      No. 55, 51 (1976)

101. L. Ma, K.M Midellemiss, G.M. Torrie, and S.G. Whittington,
      J. Chem. Soc., FaradayTrans. II, 74, 721 (1978)

102. G.M. Torrie, J. Barrett, and S.G. Whittington, J. Chem. Soc.,
      Faraday Trans.II, 75, 369 (1979)

103. P.H. Verdier and W.H. Stockmayer, J. Chem. Phys. 36, 227
      (1962)

104. P.H. Verdier, J. Chem. Phys. 45, 2118 (1966); ibid. 59,
      6119 (1973)

105. L. Monnerie and F. Gény, J. Chim. Phys. 66, 1691 (1969)

106. L. Monnerie, F. Gény, and J. Fouquet, J. Chim. Phys. 66, 1698
      (1969)

107. F. Gény and L. Monnerie, J. Chim. Phys. 66, 1708, 1872 (1969)

108. E. Dubois-Violette, F. Gény, L. Monnerie, and O. Parodi
      J. Chim. Phys. 66, 1865 (1969)

109. G. Ågren, J. Chim. Phys. 69, 329 (1972)

110. M. Doi, Polymer J. 5, 288 (1973)

111. R. Kimmich and W. Doster, J. Polymer Sci. 14, 1671 (1976)

112. T.M. Birshtein, V.N. Gridnev, Yu.Ya. Gotlib, and A.M. Skvortsov,
      Vysokomol. soyed. A 19, 1398 (1977)

113. A. Baumgärtner and K. Binder (in press).

Chapter 4

FORTRAN PROGRAMS

4.1. The MEMORY program.

MEMORY simulates the binary and ternary copolymerizations considering various reaction mechanisms. The program implements the algorithms described in the chapters 1 (MEMORY-7) and 2 (MEMORY-3,4,5 and 6).

A. <u>Input section</u>

Card 1 : NS(FORMAT I2) - the number of data sets ; NS $\leqslant$ 99.

Card 2 : MEMORY-X- it starts with the first column.

    X=3 : binary irreversible copolymerization with ultimate effect (MEMORY-3).

    X=4 : ternary irreversible copolymerization with ultimate effect (MEMORY-4).

    X=5 : binary irreversible copolymerization with penultimate effect (MEMORY-5).

    X=6 : binary reversible copolymerization taking into account the length of the terminal sequence (MEMORY-6).

    X=7 : computation of the reactivity ratios in binary irreversible copolymerization with ultimate effect (MEMORY-7).

Card 3 : it starts with the second column and ends with the column 72, with comma ; if necessary, it continues on a second, third etc. card.

```
     For X=3 :
&   LISTA 3  R1= , R2= , NA= , NB= , N= , FRMOL= , & END
     For X=4 :
&   LISTA 4  Y= , N= , R11= , R12= ,..., R32= , R33= ,
NA= , NB= , NC= , & END
     For X=5 :
&   LISTA 5  R1= , R2= , R11= , R22= , NA= , NB= , N= ,
FRMOL= , & END
     For X=6 :
&   LISTA 6  R1= , R2= , K1= , K2= , L1= , L2= , NA= ,
NB= , N= , FRMOL= , & END.
     For X=7 :
&   LISTA 7 NA= , NB= , C1= , C2= , & END
```

The symbols have following connotation :

i) R1 and R2 - reactivity ratios in binary copolymerization (r_1, r_2)

ii) R11, R12,...,R32, R33 - reactivity ratios in ternary copolymerization $(r_{12}, r_{13}, ..., r_{31}, r_{32})$; R11=R22=R33 1.

iii) R1, R2, R11, R22 - reactivity ratios in binary copolymerization
with penultimate effect (r_1, r_2, r_1', r_2').

iV) K1, K2, L1, L2 - equilibrium constants (K_1, K_2) and the minimum
length of depolymerizable terminal sequence, i.e., $(M_2)_{L_2}$.

V) Y=1, 2 or 3 indicates the first mer of the terpolymer chain (i.
e., M_1, M_2 or M_3).

Vi) FRMOL=1 (simulation is carried out in stationary conditions) or
FRMOL=2 (simulation is carried out in non-stationary conditions).

Vii)   N - polymerization degree.

Viii) (for X=3, 4, 5, 6) NA, NB, NC stands for the number of M_1, M_2
M_3 monomers (feed) per growing macro-species.

iX) (for X=7) NA and NB denotes the number of M_1 and M_2 mers in the macromolecule with the polymerization degree N.   Cl = $\left[M_2 \right] / \left[M_1 \right]$ and   C2 = $\left[M_1 \right] / \left[M_2 \right]$. $\left[\right]$ stands for the feed concentration of the monomers. The program automatically considers N = 1000.

B. Output section

i) (for X = 3,4,5,6) : - copolymer composition
- sequence distribution
- the most probable form of the copolymer

ii) (for X = 7)   : - the list of r_1, r_2 values
- copolymer composition and sequence distribution corresponding to each pair of r_1, r_2 values.

The flow chart of the MEMORY program is shown in Figure 1.

The complete listing of the program and illustrative input / output data are given below.

145

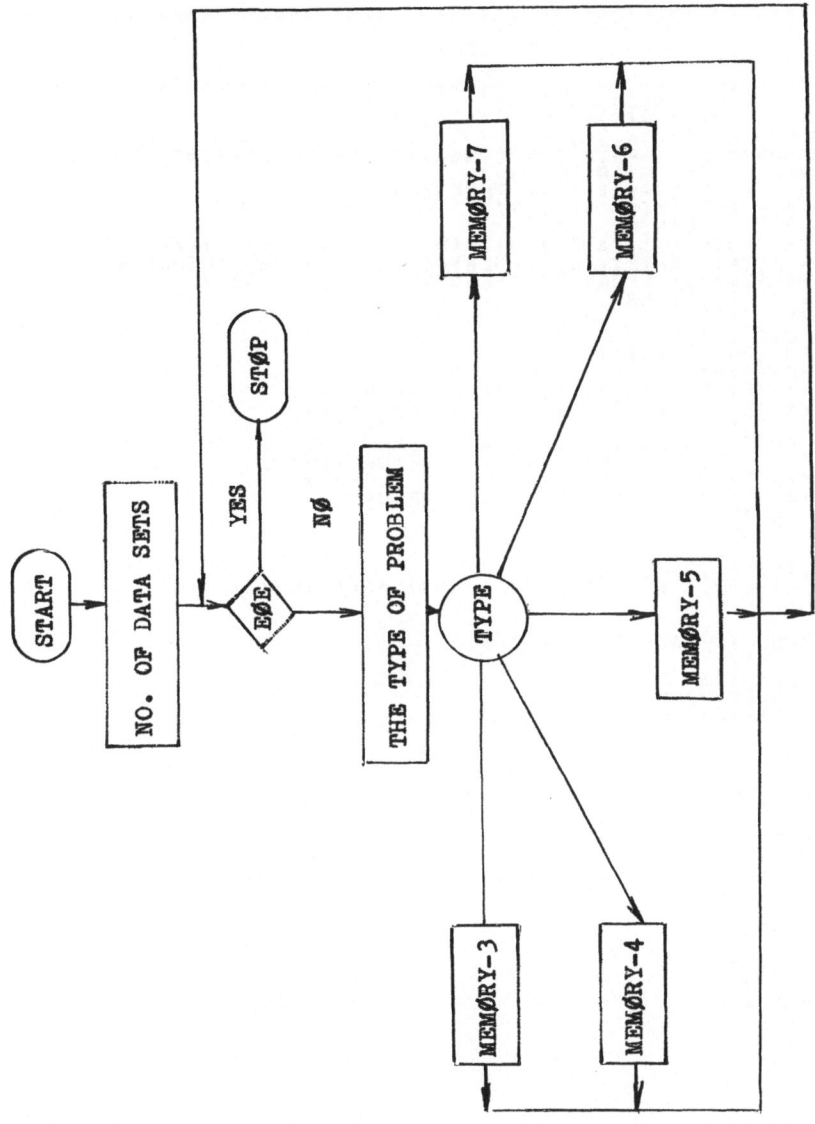

Figure 1. The flow of the MEMØRY program

```
C  ********************************************************************
C  ***                                                          **
C  ***                    M E M O R Y                           **
C  ***                                                          **
C  ********************************************************************
      DIMENSION V(3000),Q(2,2)
      COMMON/DIVID/P(3,2),R(3,3),A(3),N,MER
      COMMON/RANDM/IA,IB,X
      COMMON/KONST/NRA,NRB,NRC,NBA,NBB,NBC,NTAS,NTBS,NTCS,SW
      COMMON/TABEL/GTAB(6,16),J
      COMMON/TMERI/PKTA,PKTB,PKTC,BLK
      DOUBLE PRECISION XANTET(7)
      DOUBLE PRECISION ANTET
      DATA XANTET(1),XANTET(2),XANTET(3),XANTET(4),XANTET(5),
     *XANTET(6),XANTET(7)/'MEMORY-1','MEMORY-2','MEMORY-3',
     *'MEMORY-4','MEMORY-5','MEMORY-6','MEMORY-7'/
      READ(105,11) NS
   11 FORMAT(I1)
      DO 222 I=1,NS
      READ(105,1) ANTET
    1 FORMAT(A8)
      IF(ANTET.EQ.XANTET(1)) CALL MEMORY1(V)
      IF(ANTET.EQ.XANTET(2)) CALL MEMORY2(V)
      IF(ANTET.EQ.XANTET(3)) CALL MEMORY3(V)
      IF(ANTET.EQ.XANTET(4)) CALL MEMORY4(V)
      IF(ANTET.EQ.XANTET(5)) CALL MEMORY5(V,Q)
      IF(ANTET.EQ.XANTET(6)) CALL MEMORY6(V)
      IF(ANTET.EQ.XANTET(7)) CALL MEMORY7
      GO TO 333
  222 CONTINUE
      GO TO 444
  333 WRITE(108,2)
    2 FORMAT(////' ***ERROR THE INPUT DATA *4*////')
  444 STOP
      END
      SUBROUTINE ALEAT1(IA,IB,Z)
      CALL MAS
      IB=IA*65539
      IF(IB.GE.0) GO TO 1
      IB=IB+2147483647+1
    1 Z=IB
      Z=Z*0.4656613E-9
      IA=IB
      CALL NMAS
      RETURN
      END
      SUBROUTINE RANDOM (Z)
      DATA I/1/
      INTEGER A,X
      IF(I.EQ.0) GO TO 1
      I=0
      M=2**20
      FM=M
      X=566387
      A=2**10+3
    1 X=MOD(A*X,M)
      FX=X
      Z=FX/FM
      RETURN
      END
```

```
MASKEP    CSECT      P
          DEF        MAS,NMAS
MASQUE    DATA,4,4   X'504000000'
MAS       LDTM,13    MASQUE
          BRU        *32
NMAS      LDTM,13    MASQUE+1
          BRU        *32
          END
```

```
      SUBROUTINE MEMORY1(V)
      RETURN
      END
      SUBROUTINE MEMORY2(V)
      RETURN
      END
```

```
C       MEMORY 3
C

        DATA K1,K2,L1,L2/2*0,2*1/
        INTEGER FRMOL,N,L1,L2
        DATA A,B/'A','B'/
        DIMENSION V(3000)
        NAMELIST/LISTA3/R1,R2,NA,NB,N,FRMOL
        READ(105,LISTA3)
        WRITE(108,LISTA3)
        IA=65539
        DO 3,K=1,N
3       V(K)=0
        NNB=0
        LA=0
        LB=0
        NN=0
        NNA=0
        K=0
        IIJJ=0
        IJJ=0
        IIJ=0
        IJ=0
        IF(FRMOL.NE.1) GO TO 33
        C=NB/NA
        BETA=R2/(R2+1./C)
        GAMA1=NA/(K1+NA)
        GAMA2=NB/(K2+NB)
        ALFA=R1/(R1+C)
11      CALL MAS
        CALL ALEAT1(IA,IB,Z)
        CALL NMAS
        IJ=IJ+1
        IF(IJ.LT.6000) GO TO 11
        IF(NN.GT.N) GO TO 222
        IF(Z.GT.ALFA) GO TO 5
        IF(LA.GE.L1) GO TO 6
        K=K+1
        V(K)=A
        LA=LA+1
        NNA=NNA+1
        LB=0
        NN=NNA+NNB
        GO TO 11
5       IF(LB.GE.L2) GO TO 7
        K=K+1
        V(K)=B
        LA=0
        NNB=NNB+1
        LB=LB+1
        NN=NNA+NNB
        GO TO 44
6       IF(Z.GT.GAMA1) GO TO 11
        K=K+1
        V(K)=A

        NNA=NNA+1
        LB=0
        LA=LA+1
        NN=NNA+NNB
        GO TO 11
7       IF(Z.GT.GAMA2) GO TO 44
        K=K+1
        V(K)=B
        LA=0
        LB=LB+1
        NNB=NNB+1
        NN=NNA+NNB
        GO TO 44
8       IF(LA.GE.L1) GO TO 10
        K=K+1
        V(K)=A
        NNA=NNA+1
        LA=LA+1
        LB=0
        NN=NNA+NNB
        GO TO 11
44      CALL MAS
```

```
       CALL RANDOM(Z)
       CALL NMAS
       IIJ=IIJ+1
       IF(NN.GT.N) GO TO 222
       IF(Z.GT.BETA) GO TO 8
       IF(LB.GE.L2) GO TO 9
       K=K+1
       V(K)=B
       LA=0
       LB=LB+1
       NNB=NNB+1
       NN=NNA+NNB
       GO TO 44
9      IF(Z.GT.GAMA2) GO TO 44
       K=K+1
       V(K)=B
       LB=LB+1
       LA=0
       NNB=NNB+1
       NN=NNA+NNB
       GO TO 44
10     IF(Z.GT.GAMA1) GO TO 11
       K=K+1
       V(K)=A
       NNA=NNA+1
       LA=LA+1
       LB=0
       NN=NNA+NNB
       GO TO 11
33     IF(FRMOL.NE.2) GO TO 3333
40     C=NB/NA
       BETA=R2/(R2+1./C)
       IF(TIJ.LT.6000) GO TO 44
       GAMA1=NA/(K1+NA)
       GAMA2=NB/(K2+NB)
       ALFA=R1/(R1+C)
999    CALLMAS
       CALL ALEAT1(IA,IB,Z)
       CALL NMAS
       IJJ=IJJ+1
       IF(IJJ.LT.6000) GO TO 999
       IF(NN.GE.N) GO TO 222
       IF(Z.GT.ALFA) GO TO 50
       IF(LA.GE.L1) GO TO 60
       K=K+1

       V(K)=A
       NNA=NNA+1
       LA=LA+1
       LB=0
       NA=NA-1
       NN=NNA+NNB
       GO TO 40
50     IF(LB.GE.L2) GO TO 70
       K=K+1
       V(K)=B
       LA=0
       LB=LB+1
       NB=NB-1
       NNB=NNB+1
       NN=NNA+NNB
       GO TO 440
60     IF(Z.GT.GAMA1) GO TO 40
       K=K+1
       V(K)=A
       NA=NA-1
       NNA=NNA+1
       LB=0
       LA=LA+1
       NN=NNA+NNB
       GO TO 40
70     IF(Z.GT.GAMA2) GO TO 440
       K=K+1
       V(K)=B
       NB=NB-1
       LB=LB+1
       LA=0
       NNB=NNB+1
       NN=NNA+NNB
440    C=NB/NA
       BETA=R2/(R2+1./C)
```

```
      GAMA1=NA/(K1+NA)
      GAMA2=NB/(K2+NB)
      ALFA=R1/(R1+C)
888   CALL MAS
      CALL RANDOM(Z)
      CALL VMAS
      IIJJ=IIJJ+1
      IF(IIJJ.LT.6000) GO TO 888
      IF(NN.GT.N) GO TO 222
      IF(Z.GT.BETA) GO TO 80
      IF(LB.GE.L2) GO TO 90
      K=K+1
      V(K)=B
      LB=LB+1
      NB=NB-1
      NNB=NNB+1
      LA=0
      NN=NNA+NNB
      GO TO 440
80    IF(LA.GE.L1) GO TO 100
      K=K+1
      V(K)=A
      NA=NA-1
      NNA=NNA+1
      LB=0
      NN=NNA+NNB
      LA=LA+1
      GO TO 40
90    IF(Z.GT.GAMA2) GO TO 440
      K=K+1
      V(K)=B
      NB=NB-1
      LB=LB+1
      NNB=NNB+1
      LA=0
      NN=NNA+NNB
      GO TO 440
100   IF(Z.GT.GAMA1) GO TO 40
      K=K+1
      V(K)=A
      NA=NA-1
      NNA=NNA+1
      LB=0
      LA=LA+1
      NN=NNA+NNB
      GO TO 40
222   PRINT 55
      WRITE(108,13) (V(K),K=1,N)
55    FORMAT(////'    THE MOST PROBABLE SHAPE OF THE MOLECULE'//)
13    FORMAT((1X70A1))
      P=(100.*NNA)/N
      WRITE(108,14) P
14    FORMAT(/' P= ',F7.3)
      K=2
26    NA=0
      NB=0
      W=V(1)
      M=1
      L=2
17    IF(V(L).EQ.W) GO TO 15
      IF(M.EQ.K) GO TO 16
20    W=V(L)
      M=1
      GO TO 19
15    M=M+1
19    L=L+1
      IF(L.LE.N) GO TO 17
      IF(M.NE.K) GO TO 18
16    IF(W.EQ.A) GO TO 21
      NB=NB+1
      IF(L.GT.N) GO TO 18
      GO TO 20
21    NA=NA+1
      IF(L.GT.N) GO TO 18
      GO TO 20
18    IF(NA.EQ.0.AND.NB.EQ.0) GO TO 23
      PRINT 24,K,NA
24    FORMAT(' BLOCKS A OF LENGTH    ',I5,' = ',I5)
      PRINT 30,K,NB
30    FORMAT(' BLOCKS B OF LENGTH    ',I5,' = ',I5)
23    K=K+1
```

```
      IF(K.GT.N) GO TO 22
      GO TO 26
3333  WRITE(108,12)
12    FORMAT('FRMOL CU VALOARE NEPERMISA ')
22    STOP
      END


C
C     MEMORY 4
C
*  SEGMENT DIVID+RANDM+KONST+TABEL+TMERI
      SUBROUTINE MEMORY4(POLYM)
      INTEGER SW
      COMMON/KONST/NRA,NRB,NRC,NBA,NBB,NBC,NTAS,NTBS,NTCS,SW
      COMMON/RANDM/IA,IB,X
      COMMON/TABEL/GTAB(6,16),J
      COMMON/TMERI/PKTA,PKTB,PKTC,BLK
      COMMON/DIVID/P(3,2),R(3,3),A(3),N,MER
      DIMENSION POLYM(3000)
      NAMELIST/LISTA4/MER,N,R,A
      READ(105,LISTA4)
      WRITE(108,LISTA4)
      IA=65539
      DO 2 L=1,6000
      CALL MAS
      CALL ALEAT1(IA,IB,X)
      CALL NMAS
2     IA=IB
      CALL INIT(POLYM)
      CALL ERROR
      IF(MER.EQ.1)CALL INITA(POLYM)
      IF(MER.EQ.2)CALL INITB(POLYM)
      IF(MER.EQ.3)CALL INITC(POLYM)
      CALL PART
3     CALL ERRORP
4     CALL MAS
      CALL ALEAT1(IA,IB,X)
      CALL NMAS
      IA=IB
      CALL ERRORR
      I=I+1
      IF(I.GT.N)CALL OUT(POLYM,&8)

      GO TO (5,6,7),SW
5     CALL GENERA1(POLYM)
      GO TO 4
6     CALL GENERA2(POLYM)
      GO TO 4
7     CALL GENERA3(POLYM)
      GO TO 4
8     CALL GRUPE(POLYM)
      STOP
      END
*  SEGMENT KONST+DIVID+RANDM
      SUBROUTINE ERROR
      INTEGER SW
      COMMON/DIVID/P(3,2),R(3,3),A(3),N,MER
      COMMON/KONST/NRA,NRB,NRC,NBA,NBB,NBC,NTAS,NTBS,NTCS,SW
      COMMON/RANDM/IA,IB,X
      IF(N.LE.0)GO TO 3
      DO 1 I=1,3
      DO 1 J=1,3
      IF(R(I,J).LE.0.)GO TO 4
1     CONTINUE
      DO 2 I=1,3
      IF(A(I).LE.0.)GO TO 5
2     CONTINUE
      I=1
      IF(MER.LT.1.OR.MER.GT.3)GO TO 6
      RETURN
3     WRITE(108,900)N
      RETURN
4     WRITE(108,910)R(I,J),I,J
      RETURN
5     WRITE(108,920)A(I),I
      RETURN
6     WRITE(108,930)MER
      RETURN
```

```
  900 FORMAT(///5X,'*ERROR *    N=',I6)
  910 FORMAT(///5X,'*ERROR *    R=',F8.3,' R(',I1,',',I1,')')
  920 FORMAT(///5X,'*ERROR *    A=',F8.3,' A(',I1,',',I1,')')
  930 FORMAT(///5X,'*ERROR * MER=',I2)
      ENTRY ERRORP
      DO 7 I=1,3
      DO 7 J=1,2
      IF((P(I,J).LT.0.).OR.(P(I,J).GT.1.))GO TO 8
    7 CONTINUE
      I=1
      RETURN
    8 WRITE(108,940)P(I,J),I,J
      RETURN 1
  940 FORMAT(///5X,'*ERROR *    P=',F8.3,' P(',I1,',',I1,')')
      ENTRY ERRORR
      IF(X.LE.0..OR.X.GE.1.)GO TO 9
      RETURN
    9 WRITE(108,950)X
      RETURN
  950 FORMAT(///5X,'*ERROR *    ALEAT X=',F10.7)
      END
* SEGMENT DIVID+RANDM+KONST+TABEL+TMERI
      SUBROUTINE GENERA1(POLYM)
      INTEGER SW
      DIMENSION POLYM(10001)
      COMMON/DIVID/P(3,2),R(3,3),A(3),N,MER
      COMMON/RANDM/IA,IB,X
      COMMON/KONST/NRA,NRB,NRC,NBA,NBB,NBC,NTAS,NTBS,NTCS,SW
      COMMON/TABEL/GTAB(6,16),J
      COMMON/TMERI/PKTA,PKTB,PKTC,BLK
      IF(X.LE.P(1,1))GO TO 3
      IF(NRA.EQ.1.OR.NRA.EQ.0)GO TO 1
      NBA=NRA+1
      GTAB(1,J)=PKTA
      GTAB(2,J)=NRA
      GTAB(3,J)=NBA
      GTAB(4,J)=NTAS
      GTAB(5,J)=NTBS
      GTAB(6,J)=NTCS
      J=J+1
      IF(J.GT.14)CALL GRTAB
    1 IF(X.LE.P(1,2))GO TO 2
      POLYM(I)=PKTC
      SW=3
      NRA=0
      NRC=NRC+1
      NTCS=NTCS+1
      RETURN
    2 POLYM(I)=PKTB
      SW=2
      NRB=NRB+1
      NRA=0
      NTBS=NTBS+1
      RETURN
    3 POLYM(I)=PKTA
      SW=1
      NRA=NRA+1
      NTAS=NTAS+1
      RETURN
      END
* SEGMENT DIVID+RANDM+KONST+TABEL+TMERI
      SUBROUTINE GENERA3(POLYM)
      INTEGER SW
      DIMENSION POLYM(10001)
      COMMON/DIVID/P(3,2),R(3,3),A(3),N,MER
      COMMON/RANDM/IA,IB,X
      COMMON/KONST/NRA,NRB,NRC,NBA,NBB,NBC,NTAS,NTBS,NTCS,SW
      COMMON/TABEL/GTAB(6,16),J
      COMMON/TMERI/PKTA,PKTB,PKTC,BLK
      IF(X.GT.P(3,2))GO TO 3
      IF(NRC.EQ.1.OR.NRC.EQ.0)GO TO 1
      NBC=NBC+1
```

```
      GTAB(1,J)=PKTC
      GTAB(2,J)=NRC
      GTAB(3,J)=NBC
      GTAB(4,J)=NTAS
      GTAB(5,J)=NTBS
      GTAB(6,J)=NTCS
      J=J+1
      IF(J.GT.14)CALL GRTAB
    1 IF(X.LE.P(3,1))GO TO 2
      POLYM(I)=PKTB
      SW=2
      NRC=0
      NRB=NRB+1
      NTBS=NTBS+1
      RETURN
    2 POLYM(I)=PKTA
      SW=1
      NRC=0
      NRA=NRA+1
      NTAS=NTAS+1
      RETURN
    3 POLYM(I)=PKTC
      NRC=NRC+1
      NTCS=NTCS+1
      SW=3
      RETURN
      END
* SEGMENT DIVID+RANDM+KONST+TABEL+TMERI
      SUBROUTINE GENERA2(POLYM)
      DIMENSION POLYM(10001)
      INTEGER SW
      COMMON/DIVID/P(3,2),R(3,3),A(3),N,MER
      COMMON/RANDM/IA,IB,X
      COMMON/KONST/NRA,NRB,NRC,NBA,NBB,NBC,NTAS,NTBS,NTCS,SW
      COMMON/TABEL/GTAB(6,16),J
      COMMON/TMERI/PKTA,PKTB,PKTC,BLK
      IF(X.GT.P(2,1).AND.X.LE.P(2,2))GO TO 3
      IF(NRB.EQ.1.OR.NRB.EQ.0)GO TO 1
      NBB=NBB+1
      GTAB(1,J)=PKTB
      GTAB(2,J)=NRB
      GTAB(3,J)=NBB
      GTAB(4,J)=NTAS
      GTAB(5,J)=NTBS
      GTAB(6,J)=NTCS
      J=J+1
      IF(J.GT.14)CALL GRTAB
    1 IF(X.LE.P(2,1))GO TO 2
      POLYM(I)=PKTC
      SW=3
      NRB=0
      NRC=NRC+1
      NTCS=NTCS+1
      RETURN
    2 POLYM(I)=PKTA
      SW=1
      NRB=0
      NRA=NRA+1
      NTAS=NTAS+1
      RETURN
    3 POLYM(I)=PKTB
      NRB=NRB+1
      NTBS=NTBS+1
      SW=2
      RETURN
      END
* SEGMENT TABEL
      SUBROUTINE GRTAB
      COMMON/TABEL/GTAB(6,16),J
      WRITE(108,900)((GTAB(II,JJ),JJ=1,14),II=1,6)
      DO 1 II=1,6
      DO 1 JJ=1,16
    1 GTAB(II,JJ)=BLK
      J=1
      RETURN
```

```
  900 FORMAT(//2X,130('-')/2X,'! NATURE OF SEQUENCE    !',
     1          14('-'),A2,'   '),' !'
     2          /2X,130('-')/2X,'! LENGTH OF SEQUENCE    !',
     3          14('!',I4,'!'),' !'
     4          /2X,130('-')/2X,'! POSITION OF SEQUENCE  !',
     5          14('!',I4,'!'),' !'
     6          /2X,130('-')/2X,'! NUMBER OF A MERS      !',
     7          14('!',I4,'!'),' !'
     8          /2X,130('-')/2X,'! NUMBER OF B MERS      !',
     9          14('!',I4,'!'),' !'
     A          /2X,130('-')/2X,'! NUMBER OF C MERS      !',
     B          14('!',I4,'!'),' !'
     C          /2X,130('-'))
      END
* SEGMENT TMERI
      BLOCK DATA
      COMMON/TMERI/PKTA,PKTB,PKTC,BLK
      DATA PKTA,PKTB,PKTC,BLK/2HA-,2HB-,2HC-,2H /
      END
* SEGMENT RANDM


      SUBROUTINE ALEAT


      COMMON/RANDM/IA,IB,X
      IB=IA*65539
      IF(IB.GE.0) GO TO 1
      IB=IB+2147483647+1
    1 X=IB
      X=X*0.4656613E-9
      RETURN
      END
* SEGMENT DIVID

      SUBROUTINE PART
      COMMON/DIVID/P(3,2),R(3,3),A(3),N,MER
      DIMENSION SUMA(3)
      DO 1 I=1,3
      SUMA(I)=0.0
      DO 1 J=1,2
    1 P(I,J)=0.0
      DO 2 I=1,3
      DO 2 J=1,3
    2 SUMA(I)=SUMA(I)+A(J)/R(I,J)
      DO 4 I=1,3
      DO 3 J=1,2
    3 P(I,J)=P(I,J)+A(J)/R(I,J)/SUMA(I)
    4 P(I,2)=P(I,2)+P(I,1)
      RETURN
      END
* SEGMENT DIVID+RANDM+KONST+TABEL+TMERI

      SUBROUTINE INIT(POLYM)
      DIMENSION POLYM(10001)
      COMMON/DIVID/P(3,2),R(3,3),A(3),N,MER
      INTEGER SW
      COMMON/RANDM/IA,IB,X
      COMMON/KONST/NRA,NRB,NRC,NBA,NBB,NBC,NTAS,NTBS,NTCS,SW
      COMMON/TABEL/GTAB(6,16),J
      COMMON/TMERI/PKTA,PKTB,PKTC,BLK
      I=1
      J=1
      NRA=0
      NRB=0
      NRC=0
      NBA=0
      NBB=0
      NBC=0
      NTAS=0
      NTBS=0
```

```
      NTCS=0
      DO 1 L=1,6
      DO 1 K=1,16
    1 GTAB(L,K)=BLK
      RETURN
      ENTRY INITA
      POLYM(I)=PKTA
      NRA=1
      NTAS=1
      SW=1
      RETURN
      ENTRY INITB
      POLYM(I)=PKTB
      NRB=1
      NTBS=1
      SW=2
      RETURN
      ENTRY INITC
      POLYM(I)=PKTC
      NRC=1
      NTCS=1
      SW=3
      RETURN
      END
*  SEGMENT KONST+TABEL
      SUBROUTINE OUT(POLYM)
      DIMENSION POLYM(10001)
      COMMON/DIVID/P(3,2),R(3,3),A(3),N,MER
      COMMON/KONST/NRA,NRB,NRC,NBA,NBB,NBC,NTAS,NTBS,NTCS,SW
      COMMON/TABEL/GTAB(6,16),J
      COMMON/TMERI/PKTA,PKTB,PKTC,BLK
      IF(NRA.EQ.1.OR.NRA.EQ.0) GO TO 1
      NBA=NBA+1
      GTAB(1,J)=PKTA
      GTAB(2,J)=NRA
      GTAB(3,J)=NBA
      GTAB(4,J)=NTAS
      GTAB(6,J)=NTCS
      GTAB(5,J)=NTBS
    1 IF(NRB.EQ.1.OR.NRB.EQ.0) GO TO 2
      NBB=NBB+1
      GTAB(1,J)=PKTB
      GTAB(2,J)=NRB
      GTAB(3,J)=NBB
      GTAB(4,J)=NTAS
      GTAB(5,J)=NTBS
      GTAB(6,J)=NTCS
    2 IF(NRC.EQ.1.OR.NRC.EQ.0) GO TO 3
      NBC=NBC+1
      GTAB(1,J)=PKTC
      GTAB(2,J)=NRC
      GTAB(3,J)=NBC
      GTAB(4,J)=NTAS
      GTAB(5,J)=NTBS
      GTAB(6,J)=NTCS
    3 IF(J.GT.1) CALL GRTAB
      IF(J.EQ.1.AND.(GTAB(1,J).EQ.PKTA.OR.GTAB(1,J).EQ.PKTB.OR.
     1               GTAB(1,J).EQ.PKTC)) CALL GRTAB
      WRITE(108,900)(POLYM(I),I=1,N)
  900 FORMAT(//5X,'THE MOST PROBABLE FORM OF THE MACROMOLECULE'
     1       //5X,'===================================================='
     2       //(12X,60A2))
      RETURN 1
      END
*  SEGMENT DIVID+RANDM+TMERI
      SUBROUTINE GRUPE(POLYM)
      DIMENSION POLYM(10001)
      INTEGER SW
      COMMON/DIVID/P(3,2),R(3,3),A(3),N,MER
      COMMON/RANDM/IA,IB,X
      COMMON/TMERI/PKTA,PKTB,PKTC,BLK
      DIMENSION PA(500),PB(500),PC(500),GRUP(3,16)
      DO 19 I=1,500
      PA(I)=0.0
      PB(I)=0.0
   19 PC(I)=0.0
      DO 77 II=1,3
      DO 77 JJ=1,16
   77 GRUP(II,JJ)=BLK
```

```
      NRA=0
      NRB=0
      NRC=0
      I=1
      GO TO 8
    1 I=I+1
      IF(I.GT.N)GO TO 99
      IF(POLYM(I).EQ.PKTA.AND.KA.GE.1) GO TO 2
      IF(POLYM(I).EQ.PKTB.AND.KB.GE.1) GO TO 3
      IF(POLYM(I).EQ.PKTC.AND.KC.GE.1) GO TO 4
   99 CONTINUE
      IF(KA.GE.1) GO TO 5
      IF(KB.GE.1) GO TO 6
      IF(KC.GE.1) GO TO 7
      GO TO 1
    2 KA=KA+1
      GO TO 1
    3 KB=KB+1
      GO TO 1
    4 KC=KC+1
      GO TO 1
    5 IF(KA.GT.500) GO TO 51
      PA(KA)=PA(KA)+1
      GO TO 8
   51 PA(500)=PA(500)+1
      GO TO 8
    6 IF(KB.GT.500) GO TO 61
      PB(KB)=PB(KB)+1
      GO TO 8
   61 PB(500)=PB(500)+1
      GO TO 8
    7 IF(KC.GT.500) GO TO 71
      PC(KC)=PC(KC)+1
      GO TO 8
   71 PC(500)=PC(500)+1
      GO TO 8
    8 KA=0
      KB=0
      KC=0
      IF(I.GT.N) GO TO 9
      IF(POLYM(I).EQ.PKTA) KA=1
      IF(POLYM(I).EQ.PKTB) KB=1
      IF(POLYM(I).EQ.PKTC) KC=1
      GO TO 1
    9 I=1
      JJ=1
   12 IF(I.GE.500) GO TO 18
      SW=1
      IF(PA(I).GT.0) GO TO 13
  121 CONTINUE
      IF(PB(I).GT.0) GO TO 14
  122 CONTINUE
      IF(PC(I).GT.0) GO TO 15
      I=I+1
      GO TO 12
   13 GRUP(1,JJ)=PKTA
      GRUP(2,JJ)=I
      GRUP(3,JJ)=PA(I)
      NRA=NRA+I*PA(I)
      SW=2
      GO TO 16
   14 GRUP(1,JJ)=PKTB
      GRUP(2,JJ)=I
      GRUP(3,JJ)=PB(I)
      NRB=NRB+I*PB(I)
      SW=3
      GO TO 16
   15 GRUP(1,JJ)=PKTC
      GRUP(2,JJ)=I
      GRUP(3,JJ)=PC(I)
      NRC=NRC+I*PC(I)
      SW=1
   16 JJ=JJ+1
      IF(JJ.GT.14) GO TO 17
   10 CONTINUE
      IF(SW.EQ.2) GO TO 121
      IF(SW.EQ.3) GO TO 122
      I=I+1
      GO TO 12
```

```
   17 WRITE(108,900)((GRUP(II,JJ),JJ=1,14),II=1,3)
  900 FORMAT(//2X,130('-')/2X,'! NATURE OF SEQUENCE      !',
     1         14('!-',A2,'!   ')/2X,'!'
     2         /2X,130('-')/2X,'! LENHTH OF SEQUENCE      !',
     3         14('!-',I4,'!   ')/2X,'!'
     4         /2X,130('-')/2X,'! NUMBER OF BLOCKS        !',
     5         14('!-',I4,'!   ')/2X,'!'
     6         /2X,130('-'))
  920 FORMAT(//2X,'CONSUMED A MERS = ',I6/
     1          2X,'CONSUMED B MERS = ',I6/
     1          2X,'CONSUMED C MERS = ',I6/
     2          2X,'PERCENTAGE OF A = ',F5.2,'%'/
     2          2X,'PERCENTAGE OF B = ',F5.2,'%'/
     2          2X,'PERCENTAGE OF C = ',F5.2,'%')
        DO 11 II=1,3
        DO 11 JJ=1,16
   11 GRUP(II,JJ)=BLK
        JJ=1
        GO TO 10
   18 IF((JJ.GT.1).OR.(JJ.EQ.1.AND.(GRUP(1,1).EQ.PKTA.OR.GRUP(1,1).EQ.
     1PKTB.OR.GRUP(1,1).EQ.PKTC))WRITE(108,900)((GRUP(II,JJ),JJ=1,14),
     2II=1,3)
        PRA=FLOAT(NRA)/FLOAT(N)*100
        PRB=FLOAT(NRB)/FLOAT(N)*100
        PRC=FLOAT(NRC)/FLOAT(N)*100
        WRITE(108,920)NRA,NRB,NRC,PRA,PRB,PRC
        RETURN
        END
```

MEMORY=5

```
        SUBROUTINE MEMORY5(V,Q)
        DIMENSION V(3000),Q(2,2)
        INTEGER FRMOL,I,J,N,NA1,NA2,AI,AJ
        DATA A,B/'A','B'/
        IA=65539
        NAMELIST/LISTA5/AI,AJ,R1,R2,R11,R22,N,NA1,NA2,FRMOL
        READ(105,LISTA5)
        WRITE(108,LISTA5)
        IJ=0
        K=2
        I=AI
        J=AJ
        V(1)=I
        V(2)=J
        NN=0
        F=FLOAT(NA2)/NA1
        Q(1,1)=R1/(R1+F)
        Q(2,2)=R2/(R2+1./F)
        Q(2,1)=R11/(R11+F)
        Q(1,2)=R22/(R22+1./F)
        IF(FRMOL.NE.1) GO TO 11
    4   CALL  MAS
        CALL  ALEAT1(IA,IB,Z)
        CALL  NMAS
        IJ=IJ+1
        IF(IJ.LT.6000) GO TO 4
        NN=NN+1
        IF(NN.GT.(N-2)) GO TO 22
        K=K+1
        IF(Q(I,J).GT.Z) GO TO 3
        IF(I.EQ.1) GO TO 100
        IF(J.EQ.2) GO TO 99
        V(K)=2
        I=1
        J=2
        GO TO 4
   99   V(K)=1
        I=2
        J=1
        GO TO 4
  100   IF(I.EQ.1) IN=2
        IF(I.EQ.2) IN=1
        V(K)=IN
        J=IN
        GO TO 4
    3   IF(I.EQ.J) GO TO 200
        IF(J.EQ.1) GO TO 999
        V(K)=2
```

```
         I=2
         J=2
         GO TO 4
999      V(K)=1
         I=1
         J=1
         GO TO 4
200      IF(I.EQ.1) IN=1
         IF(I.EQ.2) IN=2
         V(K)=IN
         I=IN
         J=IN
         GO TO 4
11       IF(FRMOL.NE.2) GO TO 5
 8       CALL MAS
         CALL ALEAT1(IA,IB,Z)
         CALL NMAS
         IJ=IJ+1
         IF(IJ.LT.6000) GO TO 8
         NN=NN+1
         IF(NN.GT.(N-2)) GO TO 22
         K=K+1
         IF(Q(I,J).GT.Z) GO TO 7
         NA1=NA1-1
         IF(I.EQ.J) GO TO 300
         IF(J.EQ.2) GO TO 88
         V(K)=2
         I=1
         J=2
         GO TO 400
88       V(K)=1
         I=2
         J=1
         GO TO 400
300      IF(I.EQ.1) IN=2
         IF(I.EQ.2) IN=1
         V(K)=IN
         J=IN
400      F=FLOAT(NA2)/NA1
         Q(1,1)=R1/(R1+F)
         Q(2,2)=R2/(R2+1./F)
         Q(2,1)=R11/(R11+F)
         Q(1,2)=R22/(R22+1./F)
         GO TO 8
 7       NA2=NA2-1
         IF(I.EQ.J) GO TO 500
         IF(J.EQ.1) GO TO 888
         V(K)=2
         I=2
         J=2
         GO TO 600
888      V(K)=1
         I=1
         J=1
         GO TO 600
500      IF(I.EQ.1) IN=1
         IF(I.EQ.2) IN=2
         V(K)=IN
         I=IN
         J=IN
600      F=FLOAT(NA2)/NA1
         Q(1,1)=R1/(R1+F)
         Q(2,2)=R2/(R2+1./F)
         Q(2,1)=R11/(R11+F)
         Q(1,2)=R22/(R22+1./F)
         GO TO 8
5        WRITE(108,9) FRMOL
 9       FORMAT(' FRMOL CU VALOARE NEPERMISA:',I1)
         GO TO 25
22       NA=0
         NB=0
         K=1
12       IF(V(K).EQ.1) GO TO 10
         V(K)=B
         NB=NB+1
         GO TO 111
10       V(K)=A
         NA=NA+1
111      K=K+1
         IF(K.LE.N) GO TO 12
         PRINT 44
```

```
44      FORMAT(//' THE MOST PROBABLE SHAPE OF THE MOLECULE'//)
        WRITE(108,13) (V(K),K=1,N)
 13     FORMAT((1X70A1))
333     P=(100.*NA)/N
        WRITE(108,14) P
  14    FORMAT(//' P = ',F7.3)
        K=2
26      NA=0
        NB=0
        W=V(1)
        M=1
        L=2
17      IF(V(L).EQ.W) GO TO 15
        IF(M.EQ.K) GO TO 16
20      W=V(L)
        M=1
        GO TO 19
15      M=M+1
19      L=L+1
        IF(L.LE.N) GO TO 17
        IF(M.NE.K) GO TO 18
16      IF(W.EQ.A) GO TO 21
        NB=NB+1
        IF(L.GT.N) GO TO 18
        GO TO 20
21      NA=NA+1
        IF(L.GT.N) GO TO 18
        GO TO 20
 18     IF(NA.EQ.0.AND.NB.EQ.0) GO TO 23
        PRINT 24,K,NA
24      FORMAT(' BLOCKS A OF LENGTH ',I5,' = ',I5)
        PRINT 30,K,NB
30      FORMAT(' BLOCKS B OF LENGTH ',I5,' = ',I5)
 23     K=K+1
        IF(K.GT.N) GO TO 25
        GO TO 26
 25     STOP
        END
```

MEMORY-6

```
        SUBROUTINE MEMORY6(V)
        INTEGER FRMOL,N,L1,L2
        REAL NA,NB,K1,K2
        DATA A,B/'A','B'/
        DIMENSION V(3000)
        NAMELIST/LISTA6/R1,R2,K1,K2,NA,NB,N,L1,L2,FRMOL
        READ(105,LISTA6)
        WRITE(108,LISTA6)
        IA=65539
        DO 3,K=1,N
3       V(K)=0
        NNB=0
        LA=0
        LB=0
        NN=0
        NNA=0
        K=0
        IIJJ=0
        IJJ=0
        IIJ=0
        IJ=0
        IF(FRMOL.NE.1) GO TO 33
        C=NB/NA
        BETA=R2/(R2+1./C)
        GAMA1=NA/(K1+NA)
        GAMA2=NR/(K2+NB)
        ALFA=R1/(R1+C)
11      CALL MAS
        CALL ALEAT1(IA,IB,Z)
        CALL NMAS
        IJ=IJ+1
        IF(IJ.LT.6000) GO TO 11
        IF(NN.GT.N) GO TO 222
        IF(Z.GT.ALFA) GO TO 5
        IF(LA.GE.L1) GO TO 6
        K=K+1
        V(K)=A
        LA=LA+1
        NNA=NNA+1
```

```
        LB=0
        NN=NNA+NNB
        GO TO 11
5       IF(LB.GE.L2) GO TO 7
        K=K+1
        V(K)=B
        LA=0
        NNB=NNB+1
        LB=LB+1
        NN=NNA+NNB
        GO TO 44
6       IF(Z.GT.GAMA1) GO TO 11
        K=K+1
        V(K)=A
        NNA=NNA+1
        LB=0
        LA=LA+1
        NN=NNA+NNB
        GO TO 11
7       IF(Z.GT.GAMA2) GO TO 44
        K=K+1
        V(K)=B
        LA=0
        LB=LB+1
        NNB=NNB+1
        NN=NNA+NNB
        GO TO 44
8       IF(LA.GE.L1) GO TO 10
        K=K+1
        V(K)=A
        NNA=NNA+1
        LA=LA+1
        LB=0
        NN=NNA+NNB
        GO TO 11
44      CALL MAS
        CALL RANDOM(Z)
        CALL NMAS
        IIJ=IIJ+1
        IF(IIJ.LT.6000) GO TO 44
        IF(NN.GT.N) GO TO 222
        IF(Z.GT.BETA) GO TO 8
        IF(LB.GE.L2) GO TO 9
        K=K+1
        V(K)=B
        LA=0
        LB=LB+1
        NNB=NNB+1
        NN=NNA+NNB
        GO TO 44
9       IF(Z.GT.GAMA2) GO TO 44
        K=K+1
        V(K)=B
        LB=LB+1
        LA=0
        NNB=NNB+1
        NN=NNA+NNB
        GO TO 44
10      IF(Z.GT.GAMA1) GO TO 11
        K=K+1
        V(K)=A
        NNA=NNA+1
        LA=LA+1
        LB=0
        NN=NNA+NNB
        GO TO 11
33      IF(FRMOL.NE.2) GO TO 3333
```

```
40      C=NB/NA
        BETA=R2/(R2+1./C)
        GAMA1=NA/(K1+NA)
        GAMA2=NB/(K2+NB)
        ALFA=R1/(R1+C)
999     CALL MAS
        CALL  ALEAT1(IA,IB,Z)
        CALL  NMAS
        IJJ=IJJ+1
        IF(IJJ.LT.6000) GO TO 999
        IF(NN.GE.N) GO TO 222
        IF(Z.GT.ALFA) GO TO 50
        IF(LA.GE.L1) GO TO 60
        K=K+1
        V(K)=A
        NNA=NNA+1
        LA=LA+1
        LB=0
        NA=NA-1
        NN=NNA+NNB
        GO TO 40
50      IF(LB.GE.L2) GO TO 70
        K=K+1
        V(K)=B
        LA=0
        LB=LB+1
        NB=NB-1
        NNB=NNB+1
        NN=NNA+NNB
        GO TO 440
60      IF(Z.GT.GAMA1) GO TO 40
        K=K+1
        V(K)=A
        NA=NA-1
        NNA=NNA+1
        LB=0
        LA=LA+1
        NN=NNA+NNB
        GO TO 40
70      IF(Z.GT.GAMA2) GO TO 440
        K=K+1
        V(K)=B
        NB=NB-1
        LB=LB+1
        LA=0
        NNB=NNB+1
        NN=NNA+NNB
440     C=NB/NA
        BETA=R2/(R2+1./C)
        GAMA1=NA/(K1+NA)
        GAMA2=NB/(K2+NB)
        ALFA=R1/(R1+C)
888     CALL MAS
        CALL RANDOM(Z)
        CALL NMAS
        IIJJ=IIJJ+1
        IF(IIJJ.LT.6000) GO TO 888
        IF(NN.GT.N) GO TO 222
        IF(Z.GT.BETA) GO TO 80
        IF(LB.GE.L2) GO TO 90
        K=K+1
        V(K)=B
        LB=LB+1
        NB=NB-1
        NNB=NNB+1
        LA=0
        NN=NNA+NNB

        GO TO 440
80      IF(LA.GE.L1) GO TO 100
        K=K+1
        V(K)=A
        NA=NA-1
        NNA=NNA+1
        LB=0
        NN=NNA+NNB
        LA=LA+1
        GO TO 40
90      IF(Z.GT.GAMA2) GO TO 440
        K=K+1
        V(K)=B
```

```
        NB=NB-1
        LB=LB+1
        NNB=NNB+1
        LA=0
        NN=NNA+NNB
        GO TO 440
100     IF(Z.GT.GAMA1) GO TO 40
        K=K+1
        V(K)=A
        NA=NA-1
        NNA=NNA+1
        LB=0
        LA=LA+1
        NN=NNA+NNB
        GO TO 40
222     PRINT 55
        WRITE(108,13) (V(K),K=1,N)
55      FORMAT(////'  THE MOST PROBABLE SHAPE OF THE MOLECULE'//)
13      FORMAT((1X70A1))
        P=(100.*NNA)/N
        WRITE(108,14) P
14      FORMAT(/' P= ',F7.3)
        K=2
26      NA=0
        NB=0
        W=V(1)
        M=1
        L=2
17      IF(V(L).EQ.W) GO TO 15
        IF(M.EQ.K) GO TO 16
20      W=V(L)
        M=1
        GO TO 19
15      M=M+1
19      L=L+1
        IF(L.LE.N) GO TO 17
        IF(M.NE.K) GO TO 18
16      IF(W.EQ.A) GO TO 21
        NB=NB+1
        IF(L.GT.N) GO TO 18
        GO TO 20
21      NA=NA+1
        IF(L.GT.N) GO TO 18
        GO TO 20
18      IF(NA.EQ.0.AND.NB.EQ.0) GO TO 23
        PRINT 24,K,NA
24      FORMAT(' BLOCKS A OF LENGTH   ',I5,' = ',I5)
        PRINT 30,K,NB
30      FORMAT(' BLOCKS B OF LENGTH   ',I5,' = ',I5)
23      K=K+1
        IF(K.GT.N) GO TO 22
        GO TO 26
3333    WRITE(108,12)
12      FORMAT('FRMOL CU VALOARE NEPERMISA ')
22      STOP
        END


                        MEMORY7


        SUBROUTINE MEMORY7
        DATA N,LPA,LPB/1000,10,10/
        DATA AST,BLK/2H**,2H  /
        INTEGER BA(1000),BB(1000),M(2,90),AI,BI,LP
        DIMENSION T(2000)
        NAMELIST/LISTA7/NA,NB,C1,C2
        READ(105,LISTA7)
        WRITE(108,LISTA7)
        C=C1/C2
        IA=65539
        DO 1 I=1,6000
        CALL MAS
```

```
      CALL ALEAT1(IA,IB,X)
      CALL NMAS
1     IA=IB
      DO 2 I=1,N
      BA(I)=0
      BB(I)=0
      CALL MAS
      CALL ALEAT1(IA,IB,T(I))
      CALL NMAS
2     IA=IB
      WRITE(108,900)
      WRITE(108,910) NA,NB
      LIA=NA-2
      LSA=NA+2
      LSB=NB+2
      RN=FLOAT(N)
      DO 12 I=1,N,LPA
      A=FLOAT(I)/RN
      DO 11 J=1,N,LPB
      B=FLOAT(J)/RN
      AI=0
      BI=0
      NJ=0
      MA=0
      MB=0
      KAB=1
      DO 71 K=1,N
      GO TO (3,5),KAB
3     IF(T(K).LE.A) GO TO 4
      BI=BI+1
      NJ=NJ+1
      IF(MA.GT.0) BA(MA)=BA(MA)+1
      MA=0
      MB=MB+1
      KAB=2
      GO TO 7
4     AI=AI+1
      MA=MA+1
      GO TO 7
5     IF(T(K).LE.B) GO TO 6
      MB=MB+1
      BI=BI+1
      GO TO 7
6     AI=AI+1
      BB(MB)=BB(MB)+1
      MB=0
      MA=MA+1
      KAB=1
7     IF(AI.GT.LSA.OR.BI.GT.LSB) GO TO 33
71    CONTINUE
      IF (MA.GT.0)BA(MA)=BA(MA)+1
      IF (MB.GT.0)BB(MB)=BB(MB)+1
      IF(AI.GT.LIA.AND.AI.LT.LSA) GO TO 8
      GO TO 33
8     CONTINUE
      R1=A*C/(1.-A)
      R2=(1.-B)/(C*B)
      K=0
      DO 113 L=1,AI
      IF(BA(L).EQ.0) GO TO 113
      K=K+1
      M(1,K)=L
      M(2,K)=BA(L)
      BA(L)=0
113   CONTINUE
      IF (AI.EQ.NA) AB=AST
      WRITE(108,920)R1,R2,AB,NJ,((M(II,JJ),II=1,2),JJ=1,K)
      AB=BLK
      K=0
      DO 115 L=1,BI
      IF(BB(L).EQ.0) GO TO 115
      K=K+1
      M(1,K)=L
      M(2,K)=BB(L)
      BB(L)=0
115   CONTINUE
      WRITE(108,930)AI,BI,((M(II,JJ),II=1,2),JJ=1,K)
      GO TO 117
33    DO 112 L=1,AI
112   BA(L)=0
      DO 117 L=1,BI
117   BB(L)=0
```

```
11   CONTINUE
12   CONTINUE
     WRITE(108,900)
     RETURN
900  FORMAT(5X,125('*'))
910  FORMAT(5X,'|    R1/NA    R2/NB    NJ     |',20X,
    1'LENGTH OF SEQUENCE NA=',I3,3X,' NB=',I3,2X,/
    2 5X,125('*'))
920  FORMAT(5X,'|',2(1X,F7.2),A2,I6,' |A ',15(I3,'=',I3)/
    1(5X,'|',24X,' /A ',15(I3,'=',I3)))
930  FORMAT(5X,'|',2(2X,I6),8X,' =B ',15(I3,'=',I3)/
    1(5X,'=',24X,' IB ',15(I3,'=',I3)))
     END
```

```
&LISTA3
R1=.350000,   R2=.980000,   NA=7000.00,   NB=3000.00,   N=1000,   FRMOL=1,
&END
```

ABAAABABAABAAABAABABAAABAAABBBBABABABABABAAABBBABBBABABABABBBAAABABAAABABABAA
AAAAAAABAAABBBABABABAAAAABAAAAABBAABABABAABAAABBBABABBBAABABABABABBBABAAAAAB
ABABAAABABABBRABABABBBAAABAAABABAAABBBAAABABAAABBBABAAABBBABAAABBBAABBAABBAA
BAABAAABABABABAABBABABABABAAABABABABBBABAAABABAABAAABAAAABAABBBABAABABBA
BBABAABAABBABABAAAABAABABAABABABAABBBABAAAABBABBBBABABAAABAAABAAABBBABABAAABA
AAAAAABAABABABAAABAAABABAAABABABABBBABAAABABAAABABABABABBBAAABBBABBBABABABAABAAB
BAAAAABAABABABABAAABABARBABABAABABABAABAABABAAAABBABABBBAAABBABABBBAAABABABBBAA
BABABABAABAAAABAABBABABBBBAABABAABAAAAABAABABAAAABABABABAAABABABBBBABBBAAAA
AAAAAAAABBBAABABAABBAABAAAABABABABAABABABAABBABABABAABABAABABABABBBABABBBBA
BABABBABABAABBAAABAAABABABAABABAABABABAAAABAABABBBAABBBBAABABAABAABABABAABAA
AABAABABABAABABBAAAAABABAABBABBBABABAABAABABAABABBAABABBBAABBAAABABABAB
ABAAAABAABABBBABABABAABABBBAAAABABBABBABAABBRAABABBABABAABABAABABAAAABAAABAB
AAABAAAABAABABABAABAABABBBAAABBBBABAABBRABAAABBABABBBABBBABBBAAAAABABBBABABAB
BABABAAABBABABABABBBABAABABABABABAABBAAAABAAABABABABAABABAABABABABBBABAABABBBBAB
AAAAABAABBBABBBAAAAAAA

```
P= 57.600
BLOCKS A OF LENGTH        2 =      77
BLOCKS B OF LENGTH        2 =      55
BLOCKS A OF LENGTH        3 =      43
BLOCKS B OF LENGTH        3 =      22
BLOCKS A OF LENGTH        4 =      10
BLOCKS B OF LENGTH        4 =       3
BLOCKS A OF LENGTH        5 =       7
BLOCKS B OF LENGTH        5 =       0
BLOCKS A OF LENGTH        6 =       1
BLOCKS B OF LENGTH        6 =       0
BLOCKS A OF LENGTH        7 =       2
BLOCKS B OF LENGTH        7 =       0
BLOCKS A OF LENGTH        9 =       1
BLOCKS B OF LENGTH        9 =       0
BLOCKS A OF LENGTH       12 =       1
BLOCKS B OF LENGTH       12 =       0
```

```
&LISTA5
AI=1,  AJ=2,  R1=2.14000,  R2=.100000,  R11=.320000,  R22=1.88000,
N=1000,  NA1=5000,  NA2=5000,  FRMOL=1.
&END
```

```
ABBABBABBABBABBABBABBABBABBABABABBABBBABBBBABBBBABBBABBBAABBBABBBAAAABBABBBAAAABB
AAABABBBABBBABBBABBBABABBABBBABBBAABBABBBABABBBABBBAABABBABABBABBBABRBBABBARBBABABBA
BBABABBABAAAAAAAABBABBBABABBBAAABBBABBBABBABBBAAAABABBBAABBBBAAAABBBABABAAAAAAAA
AABBBABBBABABABABBBABABBBABABBBAAABABABBBAAAAABBBABABBBABABBBABBBAABBBAABBBAAABA
BBABABBABABABBAAAAAABBABBBAAAAAAAABBABBBABABBBABABABBBABBBABBBABBBAABBBAABBBABABA
BAABAABBABBBABBBABBBABBBABABABAAAAAAAABABBBAAABBBABABBBABABABABABBBABBBABBBABBBABA
BAAABAAAABBABBBBABBBABBBABBBAAAAAAABABABABABABABABABBBAABBBABBBABBBABBBABBBABBABA
BBAAAAAAABBBABBBABBBABBBABBBAAAABABBBABBBAAABABABABBBAABBBABBBABABAAAAABABAA
BBAAAAAAABABAAAAAAAAAABBBABBBABBBABBBAAAABBBABABAABBBAAAABBAAABAABBBAAABABABA
BAAABABBBAAAAAAAAAABABBBABABBBABBBABBBAAAABBBABBBAAABABBABBBABBBBAAABABBBABABBABB
ABBABABBBAABABABABAAAAABBBABABBBBABABBBABBBABABBBABBBABBBBBABABABABBBAABAAAAAAABBA
BABBABBBAAAABABAAABABABBBABBBAAAAAABBBABABBBAAAABABBBAAAABAABBBABABBBAAABABBBAA
AAABBBAAAAAABBBABABBBAAAAAAAABBBABBBABBBAAABABABBBAABBBAAAAAAAABAAAABAAABBBABBBABAB
ABBABBABBBABBBABBBABABABAABBBABBBABBBABBBAAABBBABBBAAABABABABABBBABBBBBABBBABABBABB
ABBABBAAAABBBAAAAAAAABB
```

```
P =  51.200
BLOCKS A OF LENGTH     2 =    22
BLOCKS B OF LENGTH     2 =   161
BLOCKS A OF LENGTH     3 =    22
BLOCKS B OF LENGTH     3 =    18
BLOCKS A OF LENGTH     4 =    13
BLOCKS B OF LENGTH     4 =     2
BLOCKS A OF LENGTH     5 =     2
BLOCKS B OF LENGTH     5 =     0
BLOCKS A OF LENGTH     6 =     6
BLOCKS B OF LENGTH     6 =     0
BLOCKS A OF LENGTH     7 =     6
BLOCKS B OF LENGTH     7 =     0
BLOCKS A OF LENGTH     8 =     3
BLOCKS B OF LENGTH     8 =     0
BLOCKS A OF LENGTH     9 =     2
BLOCKS B OF LENGTH     9 =     0
BLOCKS A OF LENGTH    12 =     1
BLOCKS B OF LENGTH    12 =     0
```

```
$LISTA6
R1=3.14000,  R2=1.20000,  K1=10.2000,  K2=5.60000,  NA=6000.00,
NB=4000.00,  N=1000,  L1=2,  L2=3,  FRMOL=1,  $END
```

```
AAAAABBBAAAAAAAAAAAABAAAABAAAAAAABBBBBRBBBABAAAAABAAABBABBBRAAAAAAAAAAAAAAA
AAAAAAAAAAAAAAAAAAAAAAAAABAABAAAABBBBBAAAAAABBBBBBABBBAAABBAAAAAAAAABBAAAAA
AABBAAAAAAAAABAAAAAAAABAAAAAAAAABAAABBBBBAAABAAAAAABBAABBBBBBAAABBBAAAAAA
AAAAABAAABAAAAAABBAAAAABBAABAAAAAAAAAAAAAABBAAAAAAAAAAAABBAAAAAAAAAAA
AAABAAABBBAAAAAAAAAAAAAAAAAAABAAAABAAAAABBBAAAAABAAAABABBAAABAAAAAB
AAAAABBBBBAAAAABAABAAAABABAAABBABBBBAAAAAAAABBBAAAAAAAAAAAAABBAAAAAABA
AAAAAAAAAAAAAAAAAAAABAAAAAAAABAABBBBAABBBAAAAABAABAAAAAAAAAAABBAAABBBAAAA
BBAAAAAAAAAAAAAAAABBBBAAABBAAAAAAABBAAAAAAAAAAAAAAAAAABBAAAAABBBBBABAA
AAAAAAABBBAAAABBAAAAABBBBBBABABBABBBBAAAAAABAAABBBAAAAAABBBBBABBABAB
AAAAAABAAAAABBAAAAABBBBAAABBBBAAAAAAAAAAABBAAAABAAABBAAAAAAAAAAAAABBAAA
ABAABBBBBBAAAAAABABAAAABRAAAAAAAAAAAAAAABBBBAAAABBBBBAABABAAAAAAAAAAA
ABAAAAAAAAAAAAAAAAAABAAAABAAAAAABAAAAAAAAABBAAAAAAAABAAAAAAAAAAABAAA
ABBBAAABABAAAAAAAAAAAAAABAAAAAAAAAAAAAAAAAAABAAAAAAABAAAAAAAAAAABABAA
ABABABBAAAAAAAAAABBABAAAAABAAABBAAAAAAAAAAAABBABAAAAAABAAABAAAAABABAAAA
AAAAAABAAAAABAAAAABB
```

```
P= 75.900
BLOCKS A OF LENGTH        2 =        9
BLOCKS B OF LENGTH        2 =       31
BLOCKS A OF LENGTH        3 =       18
BLOCKS B OF LENGTH        3 =       14
BLOCKS A OF LENGTH        4 =       13
BLOCKS B OF LENGTH        4 =       10
BLOCKS A OF LENGTH        5 =       14
BLOCKS B OF LENGTH        5 =        4
BLOCKS A OF LENGTH        6 =       10
BLOCKS B OF LENGTH        6 =        1
BLOCKS A OF LENGTH        7 =        5
BLOCKS B OF LENGTH        7 =        0
BLOCKS A OF LENGTH        8 =        6
BLOCKS B OF LENGTH        8 =        1
BLOCKS A OF LENGTH        9 =        6
BLOCKS B OF LENGTH        9 =        0
BLOCKS A OF LENGTH       10 =        2
BLOCKS B OF LENGTH       10 =        0
BLOCKS A OF LENGTH       11 =        6
BLOCKS B OF LENGTH       11 =        0
BLOCKS A OF LENGTH       12 =        2
BLOCKS B OF LENGTH       12 =        0
BLOCKS A OF LENGTH       13 =        3
BLOCKS B OF LENGTH       13 =        0
BLOCKS A OF LENGTH       14 =        1
BLOCKS B OF LENGTH       14 =        0
BLOCKS A OF LENGTH       15 =        1
BLOCKS B OF LENGTH       15 =        0
BLOCKS A OF LENGTH       17 =        2
BLOCKS B OF LENGTH       17 =        0
BLOCKS A OF LENGTH       18 =        2
BLOCKS B OF LENGTH       18 =        0
BLOCKS A OF LENGTH       19 =        1
BLOCKS B OF LENGTH       19 =        0
BLOCKS A OF LENGTH       21 =        1
BLOCKS B OF LENGTH       21 =        0
BLOCKS A OF LENGTH       22 =        1
BLOCKS B OF LENGTH       22 =        0
BLOCKS A OF LENGTH       36 =        1
BLOCKS B OF LENGTH       36 =        0
```

```
&LISTA7
NA=510,  NB=490,  C1=1.00000,  C2=1.00000,
&END
```

R1/NA	R2/NB	NJ			LENGTH OF S
.04	.04E>	477	!A	1-447	2- 28	3- 2		
509	491		-B	1-464	2- 12	3- 1		
.04	.03	481	!A	1-453	2- 26	3- 2		
511	489		-B	1-474	2- 6	3- 1		
.05	.04**	474	!A	1-440	2- 32	3- 2		
510	490		-B	1-459	2- 14	3- 1		
.06	.05	470	!A	1-433	2- 35	3- 2		
509	491		-B	1-451	2- 17	3- 2		
.08	.06	466	!A	1-423	2- 40	3- 2		
509	491		-B	1-445	2- 17	3- 4		
.08	.05**	468	!A	1-426	2- 39	3- 2		
510	490		-B	1-449	2- 16	3- 3		
.09	.06	464	!A	1-419	2- 40	3- 4		
511	489		-B	1-443	2- 17	3- 4		
.10	.10	457	!A	1-409	2- 41	3- 6		
509	491		-B	1-426	2- 28	3- 3		
.10	.09**	458	!A	1-410	2- 41	3- 6		
510	490		-B	1-429	2- 26	3- 3		
.10	.07	458	!A	1-409	2- 42	3- 6		
511	489		-B	1-430	2- 25	3- 3		
.11	.11	454	!A	1-403	2- 47	3- 4		
509	491		-B	1-421	2- 29	3- 4		
.12	.11**	454	!A	1-401	2- 50	3- 3		
510	490		-B	1-422	2- 28	3- 4		
.15	.12**	441	!A	1-380	2- 54	3- 6	4- 1
510	490		-B	1-396	2- 41	3- 4		
.18	.19	417	!A	1-340	2- 64	3- 11	4- 2
509	491		-B	1-352	2- 57	3- 7	4- 1
.18	.18**	422	!A	1-349	2- 60	3- 11	4- 2
510	490		-B	1-363	2- 51	3- 7	4- 1
.19	.19	417	!A	1-340	2- 64	3- 11	4- 2
509	491		-B	1-352	2- 57	3- 7	4- 1
.19	.18**	422	!A	1-349	2- 60	3- 11	4- 2
510	490		-B	1-363	2- 51	3- 7	4- 1
.21	.20**	409	!A	1-325	2- 70	3- 11	4- 3
510	490		-B	1-341	2- 58	3- 7	4- 3
.24	.23	400	!A	1-313	2- 69	3- 14	4- 4
509	491		-B	1-325	2- 62	3- 10	4- 3

4.2 Chain generation program

Each program has been written in an extended version of
FORTRAN IV. Program KETA generates 25600 freely jointed chains
consisting of 8 bonds (9 beads) with bond length l = 1 and a
bead diameter d (named SD in the program) = 1.

Two parameters are read from cards. The first one provides a
possibility to number the runs if repetitions are intended. By
means of the second one the random number generator is initialized.
The output contains the average values of r^i, s^i, and s^{*i} (i=2,4,6,8)
together with their standard deviations.

The following quantities are transmitted to files:

The chain coordinates to TAPE 11 by a subroutine that allows random
reading; the run number, the number of chains generated, and the
average values of r^i, s^i, s^{*i} (i = 2,4,6,8,12,16) to TAPE 20, the
current seed of the random number generator, and other data needed
for the prolongation program KETZ to TAPE 15.
Program KETZ dimerizes the chains the coordinates of which are
read from the file TAPE 11 up to chains consisting of 64 bonds.

The average values of the same quantities as in program KETA are
printed for 16-,32-, and 64-bond chains. Program KETF is a slight
modification of KETZ which works less time-consuming for long chains.
It continues the dimerizing process up to chains consisting of 1024
bonds.
Description of the library subprograms used

RANF   is a random number generator. It returns values uniformly
            distributed over the range (0,1).

RANSET   initializes seed of RANF.

RANGET   obtains current seed of RANF. The value returned may
            be passed to RANSET at a later time to regenerate the

	same sequence of random numbers.
MOVLEV	transfers consecutive words of data.
OPENMS	opens the mass storage file and informs the system that it is a random (word addressable) file.
WRITMS	transmits data from central memory to the file.
READMS	transmits data from the file to central memory
CLOSMS	closes the file
VMULFF	multiplies two matrices
RETURN	gives a file free so that it can be reused immediately.

```
      PROGRAM KETA(INPUT,OUTPUT,TAPE15=1000,TAPE11,TAPE20=1000)
      DIMENSION S(3),U(3,3),T(3,3),U1(3,3),Y(9)
      DIMENSION RQ(3,8),RQM(3,8),X(4,12),MASTER(25601)
      SD=1.00
      SDQ=SD*SD
      UG=0.5*SDQ-1.0
      M=3
      N=8
      L=25600
      MAX=L+1
      M1=M+1    $    N1=N+1
      CALL OPENMS(11,MASTER,MAX,0)
      IWA=(N+4)*4
      KV=0
      DO 5   I=1,3
      DO 5   J=1,8
5     RQM(I,J)=0.0
      X(1,1)=X(2,1)=X(3,1)=0.0
      X(1,2)=X(2,2)=X(2,3)=0.0
      X(3,2)=1.0
      READ 2005,LAUF,IR
      PRINT 2000,IR
2005  FORMAT(I3,I20)
2000  FORMAT(/////* INITIALIZATIONNUMBER *,I20)
      CALL RANSET(IR)
      DO 50 K=1,L
8     KV=KV+1
      U(1,2)=U(2,1)=U(2,3)=U(3,2)=0.0
      U(2,2)=-1.0
      U(3,3)=(1.0-UG)*RANF(IR)+UG
      U(1,1)=-U(3,3)
      U(1,3)=U(3,1)=X(1,3)=SQRT(1.0-U(3,3)**2)
      X(3,3)=1.0+U(3,3)
      X(4,3)=SQRT(X(1,3)**2+X(3,3)**2)
      DO 30 J=M1,N1
      T(3,2)=0.0
      CTH=T(3,3)=(1.0-UG)*RANF(IR)+UG
      STH=T(3,1)=SQRT(1.0-CTH*CTH)
10    XH=2.0*RANF(IR)-1.0
      YH=RANF(IR)
      XH2=XH*XH    $    YH2=YH*YH    $    R=XH2+YH2
      IF(R.GT.1.0) GOTO 10
      SFI=T(1,2)=2.0*XH*YH/R
      T(2,1)=-SFI*CTH
      T(2,3)=STH*SFI
      CFIN=T(2,2)=(YH2-XH2)/R
      T(1,1)=CTH*CFIN    $    T(1,3)=-STH*CFIN
      CALL VMULFF(U,T,3,3,3,3,3,U1,3,IER)
      DO 15 I=1,3
15    X(I,J)=X(I,J-1)+U1(I,3)
      IE=J-3
      DO 20 IA=1,IE
      JA=IE+1-IA
      XH=ABS(X(3,J)-X(3,JA))
      IF(XH.GE.SD) GOTO 20
      YH=ABS(X(1,J)-X(1,JA))
      IF(YH.GE.SD) GOTO 20
      XH2=ABS(X(2,J)-X(2,JA))
      IF(XH2.GE.SD) GOTO 20
      D=XH*XH+YH*YH+XH2*XH2
      IF(D.LT.SDQ) GOTO 8
20    CONTINUE
      CALL MOVLEV(U1,U,9)
```

```
30      CONTINUE
        DO 35  I=1,3
        X(4,I+N1)=0.0
        DO 35  J=1,3
35      X(J,I+N1)=U(J,I)
        CALL WRITMS(11,X,IWA,K,0,0)
        DO 40  I=1,3
        SUM=0.0
        DO 38  J=1,N1
38      SUM=SUM+X(I,J)
40      S(I)=SUM/N1
        SUM=0.0
        DO 42  J=1,N1
        Y(J)=(X(1,J)-S(1))**2+(X(2,J)-S(2))**2+(X(3,J)-S(3))**2
42      SUM=SUM+Y(J)
        RQ(1,1)=X(1,N1)**2+X(2,N1)**2+X(3,N1)**2
        RQ(2,1)=RQ(3,1)=SUM/N1
        DO 48  I=1,8
        IF(I.EQ.5.OR.I.EQ.7) GOTO 48
        IF(I.EQ.1) GOTO 46
        RQ(1,I)=RQ(1,1)**I
        RQ(2,I)=RQ(2,1)**I
        SUM=0.0
        DO 44 J=1,N1
44      SUM=SUM+Y(J)**I
        RQ(3,I)=SUM/N1
46      DO 47  J=1,3
47      RQM(J,I)=RQM(J,I)+RQ(J,I)
48      CONTINUE
50      CONTINUE
        WRITE(20) LAUF,L
        WRITE(20) RQM
        DO 55  J=1,8
        IF(J.EQ.5.OR.J.EQ.7)  GOTO 55
        DO  54 I=1,3
54      RQM(I,J)=RQM(I,J)/L
55      CONTINUE
        PRINT 100,LAUF,L,KV
        PRINT 110
        PRINT 120,((RQM(I,J),J=1,4),I=1,2)
        DO 60  J=1,4
        Y(J)=SQRT((RQM(3,2*J)-RQM(3,J)**2)/(N1*L-1))
        DO 60  I=1,2
60      RQ(I,J)=SQRT((RQM(I,2*J)-RQM(I,J)**2)/(L-1))
        PRINT 120,((RQ(I,J),J=1,4),I=1,2)
        PRINT 130
        PRINT 120,(RQM(3,J),J=1,4)
        PRINT 120,(Y(J),J=1,4)
        CALL RANGET(IR)    $  KZ=L/2
        IM=2
        WRITE(15) IR,KZ,M1,IM,LAUF
        PRINT 2010,IR
2010    FORMAT(//*CURRENT SEED *,I20///)
        REWIND 15
        CALL CLOSMS(11)
100     FORMAT(* DATAOF FREELY JOINTED CHAINS 1. BEAD DIAMETER/BOND
        1 LENGTH = 1.0, 1. 8 BONDS (*,I2,*. RUN)*
        2,///,*NUMBER OF CHAINS:*,5X,I6,35X,*NUMBER OF TRIALS:*,5X,I7)
110     FORMAT(///,6X,*(H!2)*,10X,*(H!4)*,10X,*(H!6)*,10X,*(H!8)*,10X,
        1 *(R!2)*,10X,*(R!4)*,10X,*(R!6)*,10X,*(R!8)*,/)
120     FORMAT(1X,8(2X,1PE10.3,3X))
130     FORMAT(///,6X,*(S!2)*,10X,*(S!4)*,10X,*(S!6)*,10X,*(S!8)*,/)
        END
```

```
      PROGRAM KETZ(OUTPUT,TAPE11,TAPE13,TAPE15=1000,TAPE20=1000)
      DIMENSION X(4,68),X1(4,36),Y(65)
      COMMON IR,KAN1,KAN3,L,IM,LAUF
      REWIND 15
      READ(15) IR,L,M,IM,LAUF
      KAN1=11
      KAN3=13
      DO 10 I=1,3
      M=2*M
      M4=M+4    $    N1=2*M+1
      N2=N1+1   $    N4=N2+2
      CALL KETTEZ(X,N4,X1,M4,Y,N1,N2)
      L=L/2
      CALL RETURN(KAN1)
      K=KAN1
      KAN1=KAN3
      KAN3=K
10    CONTINUE
      REWIND 15
      WRITE(15) IR,L ,M,IM,LAUF
      END
```

```
        SUBROUTINE KETTEZ(X,N4,X1,M4,Y,N1,N2)
        DIMENSION S(3),U(3,3),T(3,3),U1(3,3),V(3)
        DIMENSION X(4,N4),X1(4,M4),RQ(3,8),RQM(3,8)
        DIMENSION MAST1(25601),MAST3(12801),Y(N1)
        COMMON IR,KAN1,KAN3,L,IM,LAUF
        SD=1.0
        SDQ=SD*SD
        UG=0.5*SDQ-1.0
        N=N1-1      $      M=M4-4     $     M1=M+1
        M11=M-1
        MAX=2*L+1
        CALL OPENMS(KAN1,MAST1,MAX ,0)
        MAX=   L+1
        CALL OPENMS(KAN3,MAST3,MAX ,0)
        IWA=M4*4
        IWB=N4*4
        PRINT 2000,IR
2000    FORMAT(/////* INITIALIZATIONNUMBER*,I20)
        CALL RANSET(IR)
        KZA=2*L
        DO 5  I=1,3
        DO 5  J=1,8
5       RQM(I,J)=0.0
        KV=0
        DO 60 K=1,L
10      KV=KV+1
        KNEU=IFIX(KZA*RANF(IR))+1
        CALL READMS(KAN1,X,IWA,KNEU)
        DO 12 I=1,3
        DO 12 J=1,3
12      U(J,I)=X(J,M1+I)
        KNEU=IFIX(KZA*RANF(IR))+1
        CALL READMS(KAN1,X1,IWA,KNEU)
        CTH=T(3,3)=(1.0-UG)*RANF(IR)+UG
        STH=SQRT(1.0-CTH*CTH)
20      XH=2.0*RANF(IR)-1.0
        YH=RANF(IR)
        XH2=XH*XH    $    YH2=YH*YH    $    R=XH2+YH2
        IF(R.GT.1.0) GOTO 20
        SFI=2.0*XH*YH/R
        CFI=(XH2-YH2)/R
21      XH=2.0*RANF(IR)-1.0
        YH=RANF(IR)
        XH2=XH*XH    $    YH2=YH*YH    $    R=XH2+YH2
        IF(R.GT.1.0) GOTO 21
        SPSI=2.0*XH*YH/R
        CPSI=(XH2-YH2)/R
        PC=CTH*CFI
        PCS=CTH*SFI
        T(1,1)=SFI*SPSI-PC*CPSI
        T(1,2)=PC*SPSI+SFI*CPSI
        T(1,3)=STH*CFI
        T(2,1)=-PCS*CPSI-CFI*SPSI
        T(2,2)=PCS*SPSI-CFI*CPSI
        T(2,3)=STH*SFI
        T(3,1)=STH*CPSI
        T(3,2)=-STH*SPSI
        CALL VMULFF(U,T,3,3,3,3,3,U1,3,IER)
        DO 35 J=1,M
        DO 22 I=1,3
```

```
22      S(I)=X1(I,J+1)
        CALL VMULFF(U1,S,3,3,1,3,3,V,3,IER)
        DO 24 I=1,3
24      X(I,M1+J)=X(I,M1)+V(I)
        DO 28 IA=1,M
        IF(J.EQ.1 .AND. IA.EQ.1) GOTO 28
        IB=M1-IA
        XH=ABS(X(3,M1+J)-X(3,IB))
        IF(XH.GE.SD) GOTO 28
        YH=ABS(X(1,M1+J)-X(1,IB))
        IF(YH.GE.SD) GOTO 28
        XH2=ABS(X(2,M1+J)-X(2,IB))
        IF(XH2.GE.SD) GOTO 28
        D=XH*XH+YH*YH+XH2*XH2
        IF(D.LT.SDQ) GOTO 10
28      CONTINUE
35      CONTINUE

        DO 40 I=1,3
        DO 40 J=1,3
40      U(J,I)=X1(J,M1+I)
        CALL VMULFF(U1,U,3,3,3,3,3,T,3,IER)
        DO 42 I=1,3
        X(4,I+N1)=0.0
        DO 42 J=1,3
42      X(J,N1+I)=T(J,I)
        IF(N.NE.64) GOTO 410
        DO 400 J=1,N1
400     X(4,J)=SQRT(X(1,J)**2+X(2,J)**2+X(3,J)**2)
410     CONTINUE
        CALL WRITMS(KAN3,X,IWB,K,-1,0)
        DO 44 I=1,3
        SUM=0.0
        DO 43  J=1,N1
43      SUM=SUM+X(I,J)
44      S(I)=SUM/N1
        SUM=0.0
        DO 46  J=1,N1
        Y(J)=(X(1,J)-S(1))**2+(X(2,J)-S(2))**2+(X(3,J)-S(3))**2
46      SUM=SUM+Y(J)
        RQ(1,1)=X(1,N1)**2+X(2,N1)**2+X(3,N1)**2
        RQ(2,1)=RQ(3,1)=SUM/N1
        DO 52  I=1,8
        IF(I.EQ.5.OR.I.EQ.7) GOTO 52
        IF(I.EQ.1) GOTO 50
        RQ(1,I)=RQ(1,1)**I
        RQ(2,I)=RQ(2,1)**I
        SUM=0.0
        DO 48  J=1,N1
48      SUM=SUM+Y(J)**I
        RQ(3,I)=SUM/N1
50      DO 51 J=1,3
51      RQM(J,I)=RQM(J,I)+RQ(J,I)
52      CONTINUE
60      CONTINUE
        WRITE(20)RQM
        DO 65  J=1,8
        IF(J.EQ.5.OR.J.EQ.7) GOTO 65
        DO 63  I=1,3
63      RQM(I,J)=RQM(I,J)/L
65      CONTINUE
```

```
        PRINT 101
101     FORMAT(5(/))
        PRINT 100,IM,N,LAUF,L,KV
        IM=IM+1
        PRINT 110
        PRINT 120,((RQM(I,J),J=1,4),I=1,2)
        DO 70   J=1,4
        Y(J)=SQRT((RQM(3,2*J)-RQM(3,J)**2)/(N1*L-1))
        DO 70   I=1,2
70      RQ(I,J)=SQRT((RQM(I,2*J)-RQM(I,J)**2)/(L-1))
        PRINT 120,((RQ(I,J),J=1,4),I=1,2)
        PRINT 130
        PRINT 120,(RQM(3,J),J=1,4)
        PRINT 120,(Y(J),J=1,4)
        CALL RANGET(IR)     $   KZ=L/2
        PRINT 2010,IR
2010    FORMAT(//*CURRENT SEED *,I20///)
        CALL CLOSMS(KAN1)
        CALL CLOSMS(KAN3)
        RETURN
100     FORMAT(* DATA OF FREELY JOINTED CHAINS 1. BEAD DIAMETER/BOND
       1 LENGTH = 1.0,*I2,*.*,I5,* BONDS (*,I2,*. RUN)*
       2,///,*NUMBER OF CHAINS:*,5X,I6,35X,*NUMBER OF TRIALS:*,5X,I7)
110     FORMAT(///,6X,*(H!2)*,10X,*(H!4)*,10X,*(H!6)*,10X,*(H!8)*,10X,
       1 *(R!2)*,10X,*(R!4)*,10X,*(R!6)*,10X,*(R!8)*,/)
120     FORMAT(1X,8(2X,1PE10.3,3X))
130     FORMAT(///,6X,*(S!2)*,10X,*(S!4)*,10X,*(S!6)*,10X,*(S!8)*,/)
        END
```

```
        PROGRAM KETF(OUTPUT,TAPE11,TAPE13,TAPE15,TAPE20)
        DIMENSION Y(4097),IORD(4098),IFELD(4098)
        COMMON IR,KAN1,KAN3,L,IM,LAUF,X(4,4104)
        REWIND 15
        READ(15) IR,L,M,IM,LAUF
        KAN1=13
        KAN3=11
        MM=4
        IF(M.EQ.512) MM=2
        DO 10  I=1,MM
        M=2*M
        M4=M+4    $    N1=2*M+1
        N2=N1+1   $    N4=N2+2
        CALL KETTEF(N4,M4,Y,N1,IORD,N2,IFELD)
        L=L/2
        CALL RETURN(KAN1)
        K=KAN1
        KAN1=KAN3
        KAN3=K
10      CONTINUE
        REWIND 15
        WRITE(15)IR,L,M,IM,LAUF
        END
```

```
      SUBROUTINE KETTEF(N4,M4,Y,N1,IORD,N2,IFELD)
      DIMENSION RQ(3,8)RQM(3,8),S(3),U(3,3),T(3,3),U1(3,3),
     1 X1(4,2052),Y(N1),IORD(N2),IFELD(N2),
     2 V(3)
     3 ,MAST1(3201),MAST3(1601)
      COMMON IR,KAN1,KAN3,L,IM,LAUF,X(4,4104)
      EQUIVALENCE (X(1,2053),X1(1,1))
      SD=1.0
      SDQ=SD*SD
      UG=0.5*SDQ-1.0
      N=N1-1    $    M=M4-4    $    M1=M+1
      M11=M-1
      MAX=2*L+1
      CALL OPENMS(KAN1,MAST1,MAX ,0)
      MAX=L+1
      CALL OPENMS(KAN3,MAST3,MAX ,0)
      IWA=M4*4
      IWB=N4*4
      PRINT 2000,IR
2000  FORMAT(/////* INITIALIZATIONNUMBER*,I20)
      CALL RANSET(IR)
      KZA=2*L
      DO 5   I=1,3
      DO 5   J=1,8
5     RQM(I,J)=0.0
      DO 8 J=1,M1
8     IORD(2*J-1)=J
      KV=0
      DO 60 K=1,L
10    KV=KV+1
      KNEU=IFIX(KZA*RANF(IR))+1
      CALL READMS(KAN1,X ,IWA,KNEU)
      DO 12 I=1,3
      DO 12 J=1,3
12    U(J,I)=X(J,M1+I)
      DO 14 J=2,N2,2
14    IORD(J)=-1
      DO 16 J=1,N2
16    IFELD(J)=-1
      DO 18 J=1,M11
      CALL EINTR(J ,X ,M4,IORD,N2,IFELD)
18    CONTINUE
      KNEU=IFIX(KZA*RANF(IR))+1
      CALL READMS(KAN1,X1,IWA,KNEU)
      CTH=T(3,3)=(1.0-UG)*RANF(IR)+UG
      STH=SQRT(1.0-CTH*CTH)
20    XH=2.0*RANF(IR)-1.0
      YH=RANF(IR)
      XH2=XH*XH    $    YH2=YH*YH    $    R=XH2+YH2
      IF(R.GT.1.0) GOTO 20
      SFI=2.0*XH*YH/R
      CFI=(XH2-YH2)/R
21    XH=2.0*RANF(IR)-1.0
      YH=RANF(IR)
      XH2=XH*XH    $    YH2=YH*YH    $    R=XH2+YH2
      IF(R.GT.1.0) GOTO 21
      SPSI=2.0*XH*YH/R
      CPSI=(XH2-YH2)/R
      PC=CTH*CFI
      PCS=CTH*SFI
      T(1,1)=SFI*SPSI-PC*CPSI
      T(1,2)=PC*SPSI+SFI*CPSI
```

```
        T(1,3)=STH*CFI
        T(2,1)=-PCS*CPSI-CFI*SPSI
        T(2,2)=PCS*SPSI-CFI*CPSI
        T(2,3)=STH*SFI
        T(3,1)=STH*CPSI
        T(3,2)=-STH*SPSI
        CALL VMULFF(U,T,3,3,3,3,3,U1,3,IER)
        DO 35 J=1,M
        DO 22 I=1,3
22      S(I)=X1(I,J+1)
        CALL VMULFF(U1,S,3,3,1,3,3,V,3,IER)
        DO 24 I=1,3
24      X(I,M1+J)=X(I,M1)+V(I)
        X(4,M1+J)=SQRT(X(1,M1+J)**2+X(2,M1+J)**2+X(3,M1+J)**2)
        IC=IFIX(X(4,M1+J))
        IE=IC+2
        IF(IC.EQ.0) IC=1
        DO 30 I=IC,IE
        IA=IFELD(I)
        IF(IA.EQ.-1) GOTO 30
26      IB=IORD(IA)
        XH=ABS(X(3,M1+J)-X(3,IB))
        IF(XH.GE.SD) GOTO 28
        YH=ABS(X(1,M1+J)-X(1,IB))
        IF(YH.GE.SD) GOTO 28
        XH2=ABS(X(2,M1+J)-X(2,IB))
        IF(XH2.GE.SD) GOTO 28
        D=XH*XH+YH*YH+XH2*XH2
        IF(D.LT.SDQ) GOTO 10
28      IB=IORD(IA+1)
        IF(IB.EQ.-1) GOTO 30
        IA=IB
        GOTO 26
30      CONTINUE
        IF(J.EQ.1) CALL EINTR(M,X,M4,IORD,N2,IFELD)
35      CONTINUE
        DO 40 I=1,3
        DO 40 J=1,3
40      U(J,I)=X1(J,M1+I)
        CALL VMULFF(U1,U,3,3,3,3,3,T,3,IER)
        DO 42 I=1,3
        X(4,I+N1)=0.0
        DO 42 J=1,3
42      X(J,N1+I)=T(J,I)
        CALL WRITMS(KAN3,X,IWB,K,0,0)
        DO 44  I=1,3
        SUM=0.0
        DO 43  J=1,N1
43      SUM=SUM+X(I,J)
44      S(I)=SUM/N1
        SUM=0.0
        DO 46  J=1,N1
        Y(J)=X(X(1,J)-S(1))**2+(X(2,J)-S(2)**2+(X(3,J)-S(3))**2
46      SUM=SUM+Y(J)
        RQ(1,1)=X(1,N1)**2+X(2,N1)**2+X(3,N1)**2
        RQ(2,1)=RQ(3,1)=SUM/N1
        DO 52  I=1,8
        IF(I.EQ.5.OR.I.EQ.7) GOTO 52
        IF(I.EQ.1) GOTO 50
        RQ(1,I)=RQ(1,1)**I
        RQ(2,I)=RQ(2,1)**I
        SUM=0.0
```

```
         DO 48   J=1,N1
48       SUM=SUM+Y(J)**I
         RQ(3,I)=SUM/N1
50       DO 51 J=1,3
51       RQM(J,I)=RQM(J,I)+RQ(J,I)
52       CONTINUE
60       CONTINUE
         WRITE(20)RQM
         PRINT 101
101      FORMAT(5(/))
         PRINT 100,IM,N,LAUF,L,KV
         IM=IM+1
         CALL RANGET(IR)   $   KZ=L/2
         DO 65   J=1,8
         IF(J.EQ.5.OR.J.EQ.7) GOTO 65
         DO 63   I=1,3
63       RQM(I,J)=RQM(I,J)/L
65       CONTINUE
         PRINT 110
         PRINT 120,((RQM(I,J),J=1,4),I=1,2)
         DO 70   J=1,4
         Y(J)=SQRT((RQM(3,2*J)-RQM(3,J)**2)/(N1*L-1))
         DO 70   I=1,2
70       RQ(I,J)=SQRT((RQM(I,2*J)-RQM(I,J)**2)/(L-1))
         PRINT 120,((RQ(I,J),J=1,4),I=1,2)
         PRINT 130
         PRINT 120,(RQM(3,J),J=1,4)
         PRINT 120,(Y(J),J=1,4)
         PRINT 2010,IR
2010     FORMAT(//*CURRENT SEED *,I20///)
         REWIND 15
         CALL CLOSMS(KAN1)
         CALL CLOSMS(KAN3)
         RETURN
100      FORMAT(* DATA OF FREELY JOINTED CHAINS 1, BEAD DIAMETER/BOND
        1 LENGTH = 1.0,*I2,*,*,I5,* BONDS (*,I2,*, RUN)*
        2,///,*NUMBER OF CHAINS:*,5X,I6,35X,*NUMBER OF TRIALS:*,5X,I7)
110      FORMAT(///,6X,*(H!2)*,10X,*(H!4)*,10X,*(H!6)*,10X,*(H!8)*,10X,
        1 *(R!2)*,10X,*(R!4)*,10X,*(R!6)*,10X,*(R!8)*,/)
120      FORMAT(1X,8(2X,1PE10.3,3X))
130      FORMAT(///,6X,*(S!2)*,10X,*(S!4)*,10X,*(S!6)*,10X,*(S!8)*,/)
         END
```

```
      SUBROUTINE EINTR(J,X,N,IORD,M,IFELD)
      DIMENSION X(4,N),IORD(M),IFELD(M)
      IC=IFIX(X(4,J)) +1
      IZEIG=2*J-1
      IA=IFELD(IC)
      IF(IA.NE.-1) GOTO 10
      IFELD(IC)=IZEIG
      GOTO 30
10    IB=IORD(IA+1)
      IF(IB.EQ.-1) GOTO 20
      IA=IB
      GOTO 10
20    IORD(IA+1)=IZEIG
30    RETURN
      END
```

The International Journal for the Polymer Scientist covering all areas of Polymer Science

Polymer Bulletin

ISSN 0170-0839   Title No. 289

Editors:
Prof. H.-J. Cantow, Makromolekulare Chemie, Universität Freiburg; Prof. J.P. Kennedy, Dept. of Polymer Science, The University of Akron; Prof. T. Saegusa, Dept. Synthetic Chemistry, Kyoto University.

Editorial Board:
H. Batzer, Basel; S. Cesca, San Donato Milanese; K. Dušek, Prague; P.J. Flory, Stanford, CA; J. Furukawa, Tokyo; J.E. McGrath, Blacksburg, VA; H.K. Hall, Jr., Tucson, AZ; M.L. Hallensleben, Hannover; H.H. Kausch, Lausanne; T. Kelen, Budapest; M. Kryszewski, Lódź; A. Ledwith, Liverpool; R.W. Lenz, Amherst, MA; E. Maréchal, Paris; J. Meißner, Zürich; A. Nakajima, Kyoto; G. and S. Henrici Olivé, Pensacola, FL; V. Percec, Akron, OH; N.A. Platé, Moscow; C.I. Simionescu, Bucuresti; S. Sivaram, Gujarat; D.H. Solomon, Melbourne; H. Tadokoro, Osaka; M. Takayanagi, Fukuoka; I. Uematsu, Tokyo; O. Vogl, Amherst, MA; C. Wippler, Strasbourg; H. Zahn, Aachen

Editorial Assistant:
A. Heinrich, Springer-Verlag, Heidelberg

New Polymerization Reactions

1982. 26 figures. V, 172 pages
(Advances in Polymer Science, Volume 42)
ISBN 3-540-10958-7

Contents: *M. Hasegawa:* Four-Center Photopolymerization in the Crystalline State. – *D.S. Johnston:* Macrozwitterion Polymerization. – *Y. Yokoyama, H.K. Hall, Jr.:* Ring-Opening Polymerization of Atom-Bridged and Bond-Bridged Bicyclic Ethers, Acetals and Orthooesters. – *B.P. Morin, I.P. Breusova, Z.A. Rogovin:* Structural and Chemical Modifications of Cellulose Copolymerization.

Polymerizations and Polymer Properties

1982. 94 figures. V, 252 pages
(Advances in Polymer Science, Volume 43)
ISBN 3-540-11048-8

Contents: *J.P. Kennedy, V.S.C. Chang, A. Guyot:* Carbocationic Synthesis and Characterization of Polyolefins with Si-H and Si-Cl Head Groups. – *A. Fradet, E. Maréchal:* Kinetics and Mechanisms of Polyesterifications. I. Reactions of Diols with Diacids. – *E. Heidemann, W. Roth:* Synthesis and Investigation of Collagen Model Peptides. – *G.K. Elyashevich:* Thermodynamics and Kinetics of Oriental Crystallization of Flexible-Chain Polymers.

Character: between the purely archival journals of full papers and "letter journals" consisting exclusively of short communications; length of papers, 4–8 pages

High-quality papers with an international spectrum: German-speaking countries, Eastern Europe and Japan 19% each; USA 13%; France 12%; other countries 18%

Competent referee system: rejection rate 35%

Rapid publication of papers: 3-6 weeks

50 reprints of each paper free of charge

No page charge

For subscription information and sample copy write to:
Springer-Verlag, Journal Promotion Dept., P.O. Box 105280, D-6900 Heidelberg, FRG

Springer-Verlag
Berlin
Heidelberg
NewYork

G. Govil, R. V. Hosur

Conformation of Biological Molecules

New Results from NMR

1982. 92 figures. VIII, 216 pages
(NMR, Volume 20)
ISBN 3-540-10769-X

Recent developments in NMR have made it an indispensable tool in biochemistry and molecular biology. With the advent of FT-NMR techniques, it has become possible to solve problems of sensitivity, resolution and assignments for medium size molecules, and to study their conformational structure and dynamics in solutions.

Availability of labelled compounds is contributing to a wider use of NMR for biological problems. Applications in studies of multimolecular systems, dynamcis of cellular chemistry, biological control and regulation and short lived reaction intermediates at enzyme active sites have started to appear in literature and major contributions of NMR to the field of molecular biology can be expected in the future. (1247 references).

This article reviews recent trends and developments in conformational studies on biological systems using NMR. The first two chapters deal with the theoretical principles and NMR techniques used in conformational analysis. The final four chapters deal with applications to different classes of biological molecules: nucleic acids and their components, amino acids, polypeptides and proteins, saccharides and polysaccharides and organisations in biomembranes. The emphasis is on basic principles and methodology in conformational analysis of biomolecules. Detailed coverage of the literature during the period from 1972 to 1980 is provided.

NMR in Medicine

Editor: R. Damadian
1981. 77 figures. V, 174 pages
(NMR, Volume 19)
ISBN 3-540-10460-7

Contents: *R. Damadian, M. Goldsmith, L. Minkoff:* NMR Scanning. – *J. D. Weisman, L. H. Bennet, S. Maxwell, D. E. Henson:* Cancer Detection by NMR in the Living Animal. – *P. T. Beall, D. Medina, C. F. Hazlewood:* The "Systematic Effect" of Elevated Tissue and Serum Relaxation Times for Water in Animals and Humans with Cancers. – *M. Goldsmith, R. Damadian:* Proton Magnetic Resonance of Human Tissues. Further Development as a Method of Cancer Diagnosis. – *G.-J. Béné, B. Borcard, E. Hiltbrand, P. Magnin:* Medical Diagnosis by Nuclear Magnetism in the Earth Field Range. – *J. A. Koutcher, K. Zaner, R. Damadian:* ^{31}P as a Nuclear Probe for the Diagnosis and Treatment of Malignant Tissue. – *T. Glonek, C. T. Burt, M. Bárány:* ^{31}P NMR Analysis of Intact Tissue Including Several Examples of Normal and Diseased Human Muscle Determinations. – *G. N. Ling, M. Tucker:* NMR Relaxation and Water Contents in Normal Tissues and Cancer Cells.

Springer-Verlag
Berlin
Heidelberg
New York

Lecture Notes in Chemistry

Vol. 1: G. H. Wagnière, Introduction to Elementary Molecular Orbital Theory and to Semiempirical Methods. V, 109 pages. 1976.

Vol. 2: E. Clementi, Determination of Liquid Water Structure. VI, 107 pages. 1976.

Vol. 3: S. R. Niketic and K. Rasmussen, The Consistent Force Field. IX, 212 pages. 1977.

Vol. 4: A. Graovac, I. Gutman and N. Trinajstić, Topological Approach to the Chemistry of Conjugated Molecules. IX, 123 pages. 1977.

Vol. 5: R. Carbo and J. M. Riera, A General SCF Theory. XII, 210 pages. 1978.

Vol. 6: I. Hargittai, Sulphone Molecular Structures. VIII, 175 pages. 1978.

Vol. 7: Ion Cyclotron Resonance Spectrometry. Edited by H. Hartmann and K.-P. Wanczek. VI, 326 pages. 1978.

Vol. 8: E. E. Nikitin and L. Zülicke, Selected Topics of the Theory of Chemical Elementary Processes. X, 175 pages. 1978.

Vol. 9: A. Julg, Crystals as Giant Molecules. VII, 135 pages. 1978.

Vol. 10: J. Ulstrup, Charge Transfer Processes in Condensed Media. VII, 419 pages. 1979.

Vol. 11: F. A. Gianturco, The Transfer of Molecular Energies by Collision: Recent Quantum Treatments. VIII, 328 pages. 1979.

Vol. 12: The Permutation Group in Physics and Chemistry. Edited by J. Hinze. VI, 230 pages. 1979.

Vol. 13: G. Del Re et al., Electronic States of Molecules and Atom Clusters. VIII, 177 pages. 1980.

Vol. 14: E. W. Thulstrup, Aspects of the Linear and Magnetic Circular Dichroism of Planar Organic Molecules. VI, 100 pages. 1980.

Vol. 15: A.T. Balaban et al, Steric Fit in Quantitative Structure-Activity Relations. VII, 178 pages. 1980.

Vol. 16: P. Čársky and M. Urban, Ab Initio Calculations. VI, 247 pages. 1980.

Vol. 17: H. G. Hertz, Electrochemistry. X, 254 pages. 1980.

Vol. 18: S. G. Christov, Collision Theory and Statistical Theory of Chemical Reactions. XII, 322 pages. 1980.

Vol. 19: E. Clementi, Computational Aspects for Large Chemical Systems. V, 184 pages. 1980.

Vol. 20: B. Fain, Theory of Rate Processes in Condensed Media. VI, 166 pages. 1980.

Vol. 21: K. Varmuza, Pattern Recognition in Chemistry. XI, 217 pages. 1980.

Vol. 22: The Unitary Group for the Evaluation of Electronic Energy Matrix Elements. Edited J. Hinze. VI, 371 pages. 1981

Vol. 23: D. Britz, Digital Simulation in Electrochemistry. X, 120 pages. 1981.

Vol. 24: H. Primas, Chemistry, Quantum Mechanics and Reductionism. XII, 451 pages. 1981.

Vol. 25: G. P. Arrighini, Intermolecular Forces and Their Evaluation by Perturbation Theory. IX, 243 pages. 1981.

Vol. 26: S. Califano, V. Schettino and N. Neto, Lattice Dynamics of Molecular Crystals. VI, 309 pages. 1981.

Vol. 27: W. Bruns, I. Motoc, and K. F. O'Driscoll, Monte Carlo Applications in Polymer Science. V, 179 pages. 1982.